Regional Innovation Policy For Small-Medium Enterprises

Regional Innovation Policy For Small-Medium Enterprises

Edited by

Bjørn T. Asheim

Professor, Centre for Technology, Innovation and Culture, University of Oslo, Norway, and Professor, Department of Social and Economic Geography, University of Lund, Sweden

Arne Isaksen

Assistant Professor, Agder University College, and Senior Researcher, STEP Group, Norway

Claire Nauwelaers

Senior Researcher, Maastricht Economic Research Institute on Innovation and Technology (MERIT), University of Maastricht, the Netherlands

Franz Tödtling

Professor, Department of City and Regional Development, Vienna University of Economics and Business Administration, Austria

Edward Elgar
Cheltenham, UK • Northampton, MA, USA

Published by
Edward Elgar Publishing Limited
Glensanda House
Montpellier Parade
Cheltenham
Glos GL50 1UA
UK

Edward Elgar Publishing, Inc.
136 West Street
Suite 202
Northampton
Massachusetts 01060
USA

A catalogue record for this book
is available from the British Library

Library of Congress Cataloguing in Publication Data
Regional innovation policy for small-medium enterprises / edited by Bjørn
    Asheim . . . [et al.].
        p.   cm.
    Includes bibliographical references and index.
    1. Small business – Technological innovations. 2. Technological
innovations – Government policy. 3. Regional economics. I. Asheim, Bjørn
Terje.

    HD2341.R3534 2003
    338.6'42 – dc21

                                                                    2003043906

ISBN 1 84376 398 2

Printed and bound in Great Britain by MPG Books Ltd, Bodmin, Cornwall

Contents

**PART IV SMEs, INNOVATION AND REGIONS: DESIGNING
            POLICIES**

Figures

Tables

Authors

Javier Alfonso Gil is Associate Professor of Economics and Economic Development at the Universidad Autónoma de Madrid. His research interest covers the fields of development theory, institutional economics and evaluation of public policy. He has written on issues concerning economic development and institutional change.

Professor Bjørn T. Asheim is part-time professor at the Centre of Technology, Innovation and Culture, University of Oslo, Norway, and holds the chair in economic geography at the Department of Social and Economic Geography, University of Lund, Sweden. He was previously professor in human geography at the Department of Sociology and Human Geography, University of Oslo. He has also for many years been associated with the Research Council funded STEP Group (Studies in technology, innovation and economic policy) in Oslo as a part-time senior researcher and scientific advisor. He is Associate Editor of *Economic Geography and Regional Studies*, and member of the editorial board of several European scientific journals including *European Planning Studies* and *Journal of Economic Geography*. He has served as an international expert for UNCTAD, OECD and EU/DG XVI. He is well-known internationally for his research in the areas of economic and industrial geography, where his main research specializations include: comparative analyses of industrial districts and industrial clusters; SMEs and innovation policy; technological change and local economic development; globalization and endogenous regional development; regional innovation systems and learning regions. He has produced many international publications within these subjects.

Professor Poul Rind Christensen is Head of the Centre for Small Business Studies at the University of Southern Denmark. His main research areas revolve around several interrelated themes: theories on small firms' internationalization, studies of the evolving internationalization of subcontractors and evolution of regional and global supply networks, including analysis of inter-organizational competence building and collaborative rigidities as well as studies on small enterprise change management and innovation. Recently his studies and publications have centred on entrepreneurship and the entrepreneurial processes in small- and medium-sized companies. He is a member of the Danish Research Unit of Industrial Dynamics (DRUID) and board member of the research organization 'European Council for Small Business'.

Andreas Peter Cornett, Ph.D, is Associate Professor at the University of Southern Denmark. He holds the Chair of the Nordic Section of the Regional Science Association International and is member of the 'Wissenschaftlicher Beirat Akademie für Raumforschung und Landesplanung' Hannover. His areas of research are: the process of integration and the implications for peripheral regions, economic transition and the development of the international economic system, comparative regional policy (EU and national), and regional innovation systems and industrial districts. His recent publications include articles on 'Regional Cohesion in an Enlarged European Union: An Analysis of Interregional Specialization and Integration' and together with Per V. Freytag: 'Innovative relations: a business to business and business to public policy perspective'.

Gioacchino Garofoli is Full Professor in Economics at the Faculty of Economics of Insubria University (Varese, Italy), President of the Italian Section of the Regional Science Association, Coordinator of the European Doctorate on 'Economics of Production and Development', and member of the Editorial Board of the journals *Regional Studies* and *Entrepreneurship and Regional Development*. His main areas of research deal with endogenous development, industrial districts, models of local development, local innovative systems and strategic planning. He has carried out research for several Directorates of the European Commission and research missions with the World Bank and with FAO. He has published several books on the main items of research (*Endogenous Development and Southern Europe*, Avebury, Aldershot (UK); *Modelli locali di sviluppo*, Franco Angeli, Milan; *Economia del territorio*, Etas, Milan; *Industrializzazione diffusa in Lombardia*, Franco Angeli, Milan; *Desarollo Local en Europa*, with A. Vazquez Barquero, Economistas Libros, Madrid).

Arne Isaksen is assistant professor at Agder University College, Grimstad, Norway, and senior researcher at the STEP Group (Studies in technology, innovation and economic policy) in Oslo. He has a doctoral degree in economic geography from the University of Oslo. Isaksen currently conducts research on innovation systems, regional cluster development, and the role of consultancy companies in regional and industrial development. He has published several books in Norwegian, in addition to many international papers dealing with regional industrial development and regional clusters, as well as evaluation studies of innovation and regional policy instruments.

Alexander Kaufmann holds master's and doctor's degrees in economics from the Vienna University of Economics and Business Administration. After finishing his studies, he worked on several research projects concerned with technology policy. Later he returned to the Vienna University of Economics and Business Administration to become research assistant at the Institute for Urban and Regional Studies focusing on innovation theory and policy. He participated in two TSER-projects within the 4th framework programme of the European Union: 'Regional innovation systems' and 'SME policy and the regional dimension of innovation'. In 2000 he joined the division 'Systems Research Technology–Economy– Environment' of the Austrian Research Centers Seibersdorf where he is now senior researcher. His current research interests concern innovation theory – in particular the application of social systems theory on innovation research, evolutionary economics and environmental economics.

Bernard Musyck is a Regional Economist. He graduated in Economics at the Catholic University of Louvain (Belgium), received his Ph.D. from the University of Sussex (United Kingdom) and worked as a post-doctoral fellow at the Institute of Development Studies in Brighton (United Kingdom), the University of Constance (Germany) and more recently the University of Insubria (Italy). He has been teaching Economics in Cyprus and worked in consultancy for the Regional Development and Innovation team of ADE S.A. in Belgium and for the Urban and Regional Innovation Research Unit (URENIO) of the University of Thessaloniki in Greece, mainly on European projects. His research interests and publications are in regional and local development and more specifically endogenous growth, localized learning and innovation policies for small- and medium-sized firms. He is currently working for Moody's Interbank Credit Service in Cyprus as a country risk analyst.

Claire Nauwelaers is Senior Research Fellow at MERIT, the Maastricht Economic Research Institute on Innovation and Technology at the University of Maastricht, in the Netherlands. Her main areas of research and expertise revolve around the analysis of the functioning of regional and national innovation systems. She is working on policy development, analysis, evaluation, and benchmarking in the areas of Science, Technology and Innovation, for the European Commission and various European national and regional authorities. She has been notably involved in the RITTS, RIS, and Innovation Trendchart programmes of the European

Commission, and evaluated the role of Structural Funds for fostering regional innovation. She recently published books and articles dealing with policy learning in innovation, the role of institutional frameworks, new approaches to cluster policies, and innovation policies in candidate countries.

David North is Professor of Regional Development at Middlesex University Business School in London and principal consultant at the University's Centre for Enterprise & Economic Development Research (CEEDR). His interests are in the field of local and regional economic development, small business development and support policy, and regional innovation support systems. He has published the results of his work widely in journals, books and reports, some recent publications being 'The Innovativeness and Growth of Rural SMEs during the 1990s' in *Regional Studies*, 'Public Sector Support for Innovating SMEs' in *Small Business Economics*, and 'Regional Technology Initiatives: Some Insights from the English Regions' in *European Planning Studies*. With various European partners, David is currently involved in an EU Framework V study entitled 'The Future of Europe's Rural Periphery' which looks at new forms of entrepreneurship and new business formation in rural areas.

Kristian Philipsen is Associate Professor at the Department of Marketing at the University of Southern Denmark. His interest in research focuses on the vertical collaboration between companies including studies on small firm growth, small firm management and a special interest in small businesses innovation management. He has also done research on entrepreneurship and the foundation of new firms. He is for the time being on leave from the university, working as a financial manager in Greenland.

Antonia Sáez-Cala is Associate Professor of Regional Economics at the Universidad Autónoma de Madrid. Her main areas of research and expertise are regional policy and the dynamics of local industrial systems. She works mainly on the problem of innovation in local production systems and local development policy.

David Smallbone is Professor of Small and Medium Enterprises and Head of the Centre for Enterprise and Economic Development Research (CEEDR) at Middlesex University Business School in the UK. He has published widely on a range of SME-related topics, much of it with an applied policy emphasis. His current research interests include innovation in SMEs, entrepreneurship in transition economies, ethnic minority enterprise, factors influencing the survival and growth of small firms and the role of trust in business transactions in high trust and low trust environments. Together with other CEEDR colleagues, David has undertaken a variety of research-based projects for clients that include the Small Business Service in the UK, as well as for government departments at home and abroad. David also works part-time as a consultant in the field of SME policy for the OECD.

Franz Tödtling is Professor at the Department of City and Regional Development, Vienna University of Economics and Business Administration. His main research areas are innovation processes and policy, spatial innovation systems and networks as well as regional development and policy. He has been involved in a number of international research projects and co-operations funded by the European Science Foundation, the European Framework Program and the US National Science Foundation. Publications include books such as *The Governance of Innovation in Europe* (jointly with Ph. Cooke and P. Boekholt, Pinter, 2000) and a large number of articles in edited volumes and professional journals such as *Research Policy, Technovation, Journal of Technology Transfer, Papers of the Regional Science Association, Regional Studies, Environment & Planning* and *European Planning Studies.*

Antonio Vázquez-Barquero is Professor of Economics at the Universidad Autónoma de Madrid. He was Director of the Instituto del Territorio y Urbanismo in Madrid and worked as a consultant for the EU, OECD, and several Municipalities and Regional Governments. His research interests focus upon matters of economic growth, organization of production, innovation and local development policy. He has published over 50 articles and books, and in 2002 his book *Endogenous Development* was published by Routledge.

Ian Vickers is a senior researcher at the Centre for Enterprise and Economic Development Research at Middlesex University in London. He has worked on a range of research and consultancy projects, particularly on the role of innovation and new technology in regional economic development and on health and safety in small firms.

Ana Isabel Viñas-Apaolaza, from the Universidad Autónoma de Madrid of Spain, recently finished her PhD studies. During that period she was involved in research projects about innovation and SMEs, clusters of mature sectors and regional policy, such as 'SMEs and the Regional Dimension of Innovation'. Currently, her main research issues are related to policy tools for improving technological level in SMEs.

René Wintjes is Senior Research Fellow at MERIT (Maastricht Economic Research Institute on Innovation and Technology), University of Maastricht. He joined MERIT at the beginning of 1998. He studied Economic Geography at Nijmegen University and worked at the Faculty of Applied Geography and Planning, Utrecht University. He obtained his PhD at Utrecht University in 2001. The thesis deals with the regional-economic impact and localization process of foreign companies in the Netherlands. At MERIT René Wintjes is mainly involved in research on innovation policy.

Abbreviations

AICE	Technology Centre on Ceramics (Valencia)
AITEX	Technology Centre on Textiles (Valencia)
AIJU	Technology Centre on Toys (Valencia)
BIC	Business and Innovation Centre (UK)
BL	Business Link
CATT	Innovation Relay Centre (Austria)
CEEDR	Centre for Enterprise and Economic Development Research (UK)
CESFO	Centre for Small Business Studies (University of Southern Denmark)
CESTEC	SME Technological Development Centre (Italy)
CIS	Community Innovation Survey
DeC	Design Counsellor
DIC	Decentralized Industrial Creativity
DSM	De Nederlandse Staatsmijnen (The Netherlands)
DTI	Danish Technological Institutes
ERDF	European Regional Development Fund
ERP	Technology Programme (Austria)
FAZAT	Research and Training Centre for Labour and Technology, Steyr (Austria)
FIRST	Formation et Impulsion à la Recherche Scientifique et Technologique (Wallonia)
FFF	Industrial Research Promotion Fund (Austria)
FORNY	Forskningsbasert Nyskaping
GTS	Approved Technical Services (Denmark)
GTZ	Incubation and Technology Centre, Wels (Austria)
HEI	Higher Education Institution
IC	Innovation Centres (The Netherlands)
ICT	Information and Communication Technologies
IMK	Instituut voor Middel and Kleine Ondernemingen (The Netherlands)
IMPIVA	Instituto de la Mediana y Pequeña Industria en Valencia
INESCOP	Technology Centre on Footwear (Valencia)
ITC	Innovation Technology Counsellor (UK)
ITF	Innovation and Technology Fund (Austria)

KIC	Knowledge-Intensive Industrial Clustering
KIM	Knowledge Carrier in SMEs
LVC	London Lee Valley Centre
LVBIC	London Lee Valley Business and Innovation Centre
MERIT	Maastricht Economic Research Institute on Innovation and Technology
NT	Innovation and Technology programme (Northern Norway)
QSE	Qualified scientists and engineers
REGINN	Regional Innovation Programme (Norway)
RIP	Regional Innovation Premium (Austria)
RIT	Responsible Technological Innovation (Wallonia)
RTC	Regional Technology Centre
RTI	Regional Technology Initiative
RTP	Regional Technology Plan
RUSH	Regional Development Programme (Norway)
R&D	Research and Development
SMART	Small Firms Merit Award for Science and Technology (UK)
SME	Small and Medium Enterprise
SMEPOL	SME policy and the Regional Dimension of Innovation, project under the Targeted Socio-Economic Research Programme of the European Union
SPRU	Science Policy Research Unit at the University of Sussex (UK)
STEP	Studies in Technology, Innovation and Economic Policy
SWP	Software Park, Hagenberg (Austria)
S&T	Science and Technology
TEFT	Technology Diffusion from Research Institutes to SMEs (Norway)
TIC	Technological Information Centres (Denmark)
TNC	Trans-national Corporation
TSER	Targeted Socio-Economic Research Programme (European Union)
TZI	Technology Centre, Innviertel (Austria)
TZL	Technology Centre, Linz (Austria)
TZS	Technology Centre, Salzkammergut (Austria)

Foreword

This book originates from a collective work, the SMEPOL project (SME Policy and the Regional Dimension of Innovation), undertaken by seven European academic research teams, and financed by the European Commission (DG Research) under the Targeted Socio-Economic Research programme. This work was based on a comparative analysis of innovation policies for SMEs in several European regions, with the ultimate goal of identifying good practice innovation policies, of different types and in different kinds of regions.

The participants in this project are academics from: the STEP Group, Norway, acting as project coordinator; the Vienna University of Economics and Business Administration, Austria; the Southern Denmark Business School, Denmark; the Universitá degli Studi di Pavia, Italy; the University of Maastricht, The Netherlands; the Universidad Autónoma de Madrid, Spain, and the Middlesex University, United Kingdom.

The motive for such a project was straightforward. After having worked for many years on the issues of functioning and transformation of innovation systems, on the role of SMEs in regional development, or on the rationales for policy intervention in regional development and innovation, the researchers involved felt a need to forge a better link between their work and actual policy-making in those areas. The team was composed of researchers, who not only excel in their academic discipline, but also have developed good empirical knowledge of real-world policies. They all shared the wish to ensure a better flow of ideas between policy-makers and academics, both ways, recognizing the importance of integrating tacit knowledge held by the former into their own work. Quite naturally, they developed this project, trying to examine current regional, SME-targeted innovation policies in the light of contemporary knowledge on such issues.

The dialogue with policy-makers continued also after completion of the SMEPOL study, where all participating teams could fruitfully apply the results. Many signs of interest were shown to the consortium also from other researchers and policy-makers based in Europe and beyond. As a response to this broader interest we decided to publish this book, so as to diffuse the lessons learnt to a wider audience and to further the dialogue.

Drawing on the results of the SMEPOL project, this book thus endeavours to answer the following question: how should regional innovation policies targeted at SMEs, be designed and implemented in order to improve their effects in terms of raising innovativeness in firms and regions?

To respond to this broad question, the partners of the SMEPOL project analysed policy instruments in the following regions: Upper Austria, The Triangle region in southern Jutland (Denmark), Lombardy and Apulia (Italy), Northern and South-East Norway, Valencia (Spain), North London with adjacent parts of the outer metropolitan area in Hertfordshire and Essex, Wallonia (Belgium), and Limburg (in the Netherlands). Over 40 policy instruments were analysed, falling into five broad categories: direct support, technology centres, brokers, mobility schemes and upgrading of SMEs' suppliers.

In each region, the SMEPOL analysis has been carried out following a common approach. Existing data have been used, but most of the input came through interviews with a large number of firms and stakeholders of the regional innovation systems. First, SMEs' innovation patterns were investigated: nature and intensity of innovation processes, sources and motives for innovation, barriers to innovation. Second, the overall policy framework was analysed. And third, as core of the analysis, the selected policy instruments were evaluated. The policy tools were evaluated according to two main types of information, i.e. against 'lessons' from modern innovation theory, and the specific SMEs' needs for innovation support in the study regions. The book provides the reader with essential elements of this analysis.

In Part I the background and conceptual foundations for the analysis are laid down. In Chapter 1, by Smallbone, North and Vickers, the specific characteristics of SMEs with regard to innovation and their role in contemporary European economies, are presented. Complementarity between big and small firms' roles in innovation, as well as heterogeneity among the latter are notably emphasized. Chapter 2, by Asheim and Isaksen, discusses the broad concept of innovation, the interactive innovation model and regional innovation systems as a framework for analysis and policy-making. Both chapters emphasize the diversity of possible innovation paths and strategies, both at firm and region levels. The key point there is that fostering interactivity in the system is a powerful tool to improve knowledge creation and diffusion of information, leading eventually to innovation.

In Part II the specific contexts for innovation and SMEs' innovation activities in each of the study regions is put in place. Chapter 3 by Isaksen gives a general view on the national and regional economic and policy contexts, pointing to important similarities and differences between the regions, analysing the variety of approaches to innovation support and pointing to typical innovation barriers. Chapter 4, by Kaufmann and

Tödtling, goes into more detail into the innovation activities of SMEs, main problems and driving forces, in each of the study regions, with the aim of providing guidelines on which aspects policies should address in order to be effective. A key argument in this chapter is that various types of SMEs and regions face different innovation barriers and that 'one-size-fits-all' policies are no solution.

Whereas Part II focuses on the policy context, highlighting market failures and system deficits which show up as barriers to innovation processes in regions and SMEs, Part III focuses on what we may call government failures by analysing the evaluated policy instruments.

The three chapters of Part III deal in detail with the 40 innovation policy instruments analysed under the SMEPOL project. Chapter 5, by Garofoli and Musyck, provides a detailed overview of policy instruments, according to a simple typology. Chapter 6, by Alfonso Gil, Sáez Cala, Vázquez Barquero and Viñas Apaolaza, summarizes the lessons learned from the evaluations of the instruments, distinguishing between short-term results and longer-term impacts for the various types of tools. By means of a benchmarking process, the most successful instruments are pointed out, within each category of tools. Chapter 7, by Christensen, Cornett and Philipsen, deals with the very important question of policy organization, discussing the multifacet notion of coherence of innovation policy. How to reconcile the need for, on the one hand, responsiveness and flexibility, and on the other hand, simplicity, transparency and stability in policy, forms a core question in this chapter.

Part IV contains the concluding Chapter 8 by Nauwelaers and Wintjes. This chapter presents a synthesized view on the state-of-the-art of SME-oriented innovation policies at regional level, as compared to the theoretical and empirical knowledge about SMEs, innovation and the regions. From this, it draws the overall conclusion that a new paradigm is needed for innovation policies, in which fostering interactive learning within the firms and within the regions plays the central role. Concrete suggestions for improving or reorienting individual policy tools, along the lines of this new paradigm, are delivered. Finally, the chapter calls for an interactive mode of policy intervention, as well as for more 'policy intelligence' in this complex field.

The editors wish to thank all those who have contributed to this project, beyond the authors mentioned in the book: first, the European Commission, DG Research, who funded most of the project from which this book originates. Complementary funding from the Norwegian Ministry of Local Affairs and Regional Development, for the Norwegian part of the project, is also acknowledged. Valuable comments on the final version of the project were received from Michael Storper, in charge of the evaluation of the project by the Commission. And, last but not least, all those SMEs' managers and policy-makers who devoted time to answering our questions during the analysis phase in the regions are also warmly thanked.

Bjørn T. Asheim
Arne Isaksen
Claire Nauwelaers
and Franz Tödtling

PART I

SMEs, INNOVATION AND REGIONS:
CONCEPTUAL BACKGROUND

1. The role and characteristics of SMEs in innovation

David Smallbone, David North and Ian Vickers

1.1 INTRODUCTION

The main objective of the SMEPOL project is to undertake comparative analysis of existing policies and programmes to encourage and support innovation in small and medium enterprises (SMEs) in order to establish good practice. Since SMEs have been increasingly recognized by policy-makers as a target for policy intervention, an important part of the context for the SMEPOL project is the role of SMEs in contemporary European economies. Another is the distinctive characteristics of SMEs with respect to innovation, since these are likely to influence the rationale for policy intervention designed to support innovation in them, as well as any barriers they face in achieving their innovative potential. It may be suggested that one of the key underlying aims of policy intervention in this respect should be to help firms to overcome any size-related barriers that may limit their ability to achieve their potential contribution to the innovative capability of a local, regional or national economy.

Although there are variations between policy programmes (within as well as between countries) in the eligibility criteria that are applied, the SME definition used in the SMEPOL project is that currently used by the EC for statistical purposes: very small enterprises, with less than 10 employees; small enterprises with 10–49 employees and a turnover of less than €7m; and medium enterprises with 50–249 employees and an annual turnover of less than €40m.

1.2 THE ROLE OF SMEs IN CONTEMPORARY EUROPEAN ECONOMIES

One aspect of the structural change that has been affecting most European economies in the last two decades is a growth in the number of small firms and an increase in their relative importance. Whilst the starting points and pace of change may vary between countries, the underlying trend is consistent, which helps to explain the increasing attention to the needs of the SME sector by policy-makers. In the more industrialized countries, this re-emergence of the small firm sector followed a period of increasing concentration, particularly in manufacturing. This reflected a previous emphasis on scale economies associated with the mass production of standardized products, market expansion to minimize costs at optimal plant sizes and an extensive division of labour. The reversal of this trend towards increasing size of enterprises and establishments in the 1970s saw the share of small enterprises start to grow, particularly in terms of employment (Sengenberger et al. 1990).

During the 1980s, the apparent ability of small enterprises to create jobs at a time when many large firms were shedding labour attracted the attention of policy-makers in many countries. Whilst the magnitude of the increase varied considerably from country to country, as well as between sectors, Sengenberger et al. concluded that the increasing share of small enterprises (<100 employees) in total employment represented a reversal of a decline in the share of the small sector that had existed for several decades previously. However, whilst emphasizing that data limitations (particularly with respect to the availability of comparable time series) limit the scope for direct comparison, Storey has suggested that the evidence available to support the Sengenberger hypothesis concerning the increasing contribution of small firms to employment is less convincing when the UK is excluded (Storey 1994). In this regard, it must be remembered that the UK started from a lower base of new firm formation and of small business activity in the 1970s than most other European countries.

Nevertheless, following the publication of the Birch Report (Birch 1979) which showed that between 1969–76, SMEs accounted for approximately 80 per cent of net employment growth in the US, the contribution of SMEs to employment generation became a focus of attention of policy-makers at the local, regional and national levels in a number of European countries (Storey 1988). Although later empirical results were less dramatic than the earlier Birch findings, various studies in market economies that include a variety of time spans and both recessionary and non-recessionary conditions, reinforced the original message with respect to the disproportionate contribution of small enterprises to new job creation (e.g. Storey and Johnson 1987; Sengenberger et al. 1990; Lageman et al. 1999).

Moreover, although job losses tend to increase during recessionary periods while the number of job gains decreases, SME performance in terms of employment has been shown to be relatively stable over the economic cycle in comparison with larger firms (Davis and Haltwanger 1992; EIM 1994; Fendel and Frenkel 1998; Schmidt 1995), particularly that of the smallest enterprises. For example, in EU countries at the beginning of the recession in 1991, employment in micro-enterprises continued to increase whilst that in small and medium firms remained unchanged and that in large enterprises declined. However, during the recovery in 1994–5, large enterprises were the first to increase their employment (EIM 1997). It would appear that employment in large enterprises is more vulnerable to cyclical effects than employment in SMEs. During recessionary periods, SMEs are able to partly offset job losses in large firms while during recovery periods employment growth is concentrated in large enterprises (EIM op. cit.).

In seeking to explain the re-emergence of the small firm sector, David Storey (1994) has distinguished between supply- and demand-side influences. On the supply side, technological change has contributed to the development of new products and services and the emergence of new knowledge-based activities, which has in turn created opportunities for new and small companies. In some sectors, such as printing, technological change has contributed to a reduction in the minimum optimum scale of production, enabling smaller firms to compete more actively with larger enterprises. In other cases, such as business services, the growth in the number of small firms is based on the new 'commodity' of information (Storey op. cit.). Increasing opportunities for smaller firms have also resulted from the fragmentation and cost reduction strategies of larger enterprises which have contributed to an increase in outsourcing and subcontracting out. Whilst it is difficult to estimate the extent to which the increase in either the number of small firms, or the people they employ, is a result of externalization strategies of large firms, it is clear that the growth of the small firms sector, during the 1970s and 1980s particularly, reflected changes in the nature of the relationships between firms of different sizes.

The growth in the number and role of small firms is associated with structural shifts between manufacturing and services, alongside an increase in the demand for business services. Rising real incomes have contributed to a growth in consumer demand for a more varied range of products and services, thereby creating niche opportunities which small firms are ideally suited to serve. There has also been an increase in the supply of potential entrepreneurs, influenced during some periods by recession-push factors that can lead to so-called 'enforced entrepreneurship'. Finally, government policies, which, in Storey's analysis, include privatization, deregulation and taxation policy as well as direct support measures designed to promote or assist small business development, have also contributed. Whilst the precise nature and extent of the commitment to SME policy varies between European countries, most EU member states attempt to encourage the development of SMEs through some combination of policies designed to improve the environment for enterprise establishment and development (such as through deregulation, liberalization, tax reforms) and direct support measures (such as financial schemes or tax relief) (EIM 1994).

It is clear that the re-emergence of the small firms sector in Europe's mature market economies is associated with a change in the nature of the relationships between firms of different sizes, which has involved a number of processes. Outsourcing and contracting out by large enterprises create business opportunities for small firms which, in some cases, can contribute to the development of clustering in which a few large enterprises act as final producers, and as customers for their small firm suppliers. Small firms in this instance can contribute to regional competitiveness through their supply role to large firms and to regional innovation capacity through the dynamic nature of their inter-relationships with large firms. Moreover, from an economic development perspective, the efficiency of the local supply system and the ability of SMEs to develop linkages with larger firms affect not just regional competitiveness but also the spread effects associated with any expansion of leading firms within the region. In other cases, regional competitiveness may be based on the interdependence between SMEs rather than on the strategies and actions of individual firms or a dependence of SMEs on larger firms. Whilst it may be argued that the social conditions within which such integrated production systems develop are highly specific, the potential advantages of co-operation in terms of external economies of scale between SMEs, between SMEs and large firms and between SMEs and external agencies makes regions where such networks exist of considerable policy interest, even if the importance of local economic networks for the majority of small firms is the exception rather than the rule (Curran and Blackburn 1994).

Differences between EU countries with respect to the relative importance of the SME sector can be demonstrated with respect to Table 1.1 which shows the share of total employment contributed by firms of different sizes. It shows that there is considerable variation in the role of SMEs in total employment between the seven countries featuring in the SMEPOL project. On the one hand, in the UK and Netherlands, SMEs represent a smaller share of total employment than the EU average (66 per cent), whilst on the other hand, in Italy, Spain, Norway and Denmark, SMEs make an above-average contribution. Indeed, in Spain and Italy, only 20 per cent of total employment is represented by large enterprises.

It is clear that there are a variety of potential contributions that SMEs can make to economic development at the regional and national levels, that include employment generation, acting as suppliers to large companies and contributing to a more diversified economic structure through the development of new activities, particularly through new firm formation in service activities. In the following section, we consider their potential contribution with respect to a region's innovative capability.

Table 1.1   Percentage employment share 1996 by size class[a]

Region	Very small	Small	Medium	Large	All	Total employment (1000)
Austria	25	19	21	35	100%	2 470
Belgium	48	14	11	27	100%	3 835
Denmark	30	22	18	30	100%	1 590
Finland	23	16	17	44	100%	1 030
France	32	19	15	34	100%	15 310
Germany	24	20	14	43	100%	29 090
Greece	47	18	14	21	100%	1 585
Ireland	18	16	14	51	100%	840
Italy	48	21	11	20	100%	14 040
Lux/burg	19	26	29	29	100%	155
Netherlands	26	19	15	40	100%	5 295
Norway	32	21	18	29	100%	1 045
Portugal	38	23	18	21	100%	2 800
Spain	47	19	12	21	100%	10 910
Sweden	25	17	16	41	100%	2 030
UK	31	16	12	41	100%	20 420
EU	33	19	14	34	100%	111 405

Note: [a] Totals may not add up exactly due to rounding off.

Source: EIM (1997) 'The European Observatory for SMEs: Fifth Annual Report', Table A2, p.305

1.3   THE CONTRIBUTION OF SMEs TO INNOVATION

There has been much debate in the literature about the relative contribution of firms of different sizes to innovation. For example, based on an analysis of the size distribution of innovating firms in the UK between 1945–83, Pavitt et al. (1987) concluded that small firms are more likely to introduce new innovations than larger firms because they have less commitment to existing practices and products than larger enterprises. Innovative SMEs can also be important in developing radically new innovation through their contribution to maintaining technological diversity, since large firms typically innovate incrementally within existing technological trajectories or paradigms. Small firms can be particularly important to high-tech sectors as sources of significant innovations, as has particularly been the case in the United States. Acs and Audretsch (1990) provided further empirical support for the disproportionate contribution of SMEs to innovation, based on US data.

On the other hand, evidence from the second Community Innovation Survey (CIS) suggests that across all sectors SMEs made fewer innovations (in terms of introducing technologically new or improved products or pro- cesses) between 1994–96 than large enterprises. Across Europe (excluding Spain, where the results are not yet available) the proportion of innovating firms by size class varied from 73 per cent of large firms, 49 per cent of medium-sized firms to 37 per cent of small firms. Moreover, the gap is wider when only 'novel' or 'radical' innovations are considered. For example, in the UK, large firms were three times more likely to be novel innovators than SMEs, both in manufacturing and services (Craggs and Jones 1998). Evidence from another recent study, which involved a re-examination of evidence from the Science Policy Research Unit's (SPRU) Innovations Database, also challenged earlier findings. The results showed that the largest enterprises have consistently been a disproportionately important source of innovation in the manufacturing sector in the UK (Tether et al. 1997).

Whilst this debate is likely to continue, not least because different studies typically use different definitions of what constitutes innovation as well as different databases to examine it, the point to stress is that both large and small firms play important, if different, roles in innovation. In other words, viewed across all sectors and types of innovation there is no optimal firm size from the point of view of innovation and that dynamic complementar- ities exist between large and small enterprises (Tether et al. 1997; Rothwell 1983; Pavitt et al. 1989). In this respect, it can be argued that whilst it is questionable whether SMEs are more or less likely to introduce fundamen- tally new (or novel) innovations than large firms (Storey and Sykes 1996), they do have a greater ability to make more incremental innovations as a result of the niche role which they often perform (Storey 1994). Further- more, small firms can be a source of important innovations, which are sub- sequently commercialized by large firms, hence creating employment, which ultimately end up in large firms.

It is important to recognize the heterogeneity that exists within the SME sector and that this has implications for the contribution of different types of SME with respect to innovation. On the one hand, there are many con- servatively managed traditional SMEs, operating in niches that are rela- tively untouched by technological change, where innovation is not an issue for managers. On the other hand, there are highly innovative new technol- ogy based firms whose knowledge base makes them potential world leaders in a specific field. Perhaps the point to stress is that as markets become increasingly internationalized, a lack of innovation (broadly defined) means that SMEs performing the former type of role look increasingly vulnerable, particularly in mature market economies. As a result, it is the importance of innovation to the competitiveness of individual firms (of all sizes) that needs to be stressed, with the need for policy to encourage and support that process. The qualitative differences that exist between SMEs with respect to innovation, suggest that a segmented approach to innovation support policy is essential if it is to be effective.

One of the factors helping to explain the heterogeneity that exists between SMEs with respect to innovation is the sectoral context, which has led some authors to attempt to classify the various roles that small firms play in relation to innovative activities. Rizzoni (1991) for example, has produced a six-fold classification based on the role of small manufacturing firms with respect to technological innovation, in which the sectoral dimension is very important. These ranged on the one hand from 'static' small firms in the sense of being largely uninvolved with innovation and showing a degree of conservatism and inefficiency, and 'traditional' small firms which play a more active role in the diffusion of innovation, to 'new technology' based small firms at the other extreme, where small firms play an important role in the introduction of significant new technologies. Rizzoni's taxonomy can be used to identify the variety of roles that small firms can play in technological change, thus emphasizing the heterogeneity that is a recurrent characteristic of the SME sector.

Similarly, Hassink (1996) has developed a typology, initially put forward by others (McKinsey & Company 1987; Rothwell 1991), to distinguish between (i) technology-driven SMEs which need to keep abreast of leading-edge technologies; (ii) technology-following SMEs where technology though important does not have to be the most advanced available; and (iii) technology-indifferent SMEs, which are essentially craft firms, and which rarely invest in new technological equipment. Any assessment of firms' support needs should be looked at in relation to the technology base of SMEs as well as the differences that can be expected between firms in different sectors. At the same time, there are also dangers in adopting an overly deterministic perspective in this respect.

The point to stress is that the role which different types of innovation play in the competitiveness of SMEs can vary considerably between sectors, which in turn has implications for what innovation means in practice and thus for the type of policy support that is appropriate to particular firms. For example, in some sectors (such as branches of engineering), the development of innovative, proprietary products which are new to the industry is likely to be an important means which firms use to seek competitive advantage. As a result, firms in these sectors that are seeking to differentiate their products from those of their competitors in order to compete on the basis of non-price advantages, have an incentive to seek innovative product development, in order to create and/or maintain specialized niches for proprietary products. The process of innovation in such firms may require a substantial resource commitment at the pre-launch stage that can be particularly demanding for new and young companies. The support needs in such cases often focus on financing and the need for external technical advice in order to extend the firm's internal resource base.

By contrast, in a sector such as food processing, competitiveness may depend more upon SMEs being able to maintain a high level of flexibility to customer demands that involves them developing slightly modified versions of existing products, or new ways of packaging them. In other words, innovative SMEs typically rely on incremental changes to their product portfolio and the way it is presented to customers rather than making more fundamental innovations or radical changes to the product range or the way that products are produced. This does not mean that SMEs cannot achieve more significant innovations in the food industry but simply that these are likely to be relatively infrequent. The main implication for support needs emphasizes the importance of marketing and in particular the nature and extent of the firm's ability to develop value-added links with its major customers.

An assessment of the performance of SMEs in terms of innovation is undoubtedly influenced by the definition of innovation that is used, as well as between different types of SME. Adopting a Porterian view, innovation is an attempt 'to create competitive advantage by perceiving or discovering new and better ways of competing in an industry and bringing them to market' (Porter 1990). Innovation is broadly defined to include both improvements in technology and better methods or ways of doing things, which can be manifested in product changes, process changes, new approaches to marketing and/or new forms of distribution. An approach to innovation which emphasizes the application of ideas and methods that are new to the firm inevitably means that much innovation in practice can appear rather mundane and incremental rather than radical from an industry perspective, depending upon an accumulation of small insights rather than on major breakthroughs. Moreover, as Porter stated, innovation can result from informal organizational learning as much as from formal research and development, which is a view that also emerges from much of the recent innovation literature which conceives innovation as an interactive, non-linear process. The focus of Bessant et al. (1994) on 'continuous improvement' is also relevant here as an important complement to radical, step-change forms of innovation, particularly since this type of innovation often results from an essentially internal process of 'learning by doing'.

Since the number of firms which are likely to be truly innovative in a technological sense is fairly small, an emphasis on a broad concept of innovation is more appropriate as far as the majority of SMEs, which is reflected in the approach to innovation of some policy programmes evaluated in the SMEPOL project (e.g. Business Link in the UK).

Whilst it is the firm that has been adopted as the main focus of investigation in the SMEPOL project, a distinction between the organization and the entrepreneur does have potential implications from a policy perspective. Since individual entrepreneurs may be involved in a number of business enterprises, an individual firm that appears non-innovative could be part of a highly innovating group of companies within an individual entrepreneur's portfolio. As Scott and Rosa (1996) pointed out, portfolio entrepreneurship can result in a very different picture of growth (or innovation) when the unit of analysis is shifted from individual firms to entrepreneurs. However, the greater difficulties in implementing such an approach is an important practical constraint, both for researchers and policy-makers.

1.4 THE DISTINCTIVENESS OF SMEs AT THE MICRO LEVEL AND THEIR SUPPORT NEEDS

Despite the variations that can exist empirically between the extent of innovation in SMEs, both between and within sectors, there are certain size-related characteristics of SMEs at the micro level that can contribute to the shaping of strategic activity and underlying management actions affecting the innovation process which have potential implications both for support needs and for the way in which those needs are addressed. Whilst it must be recognized that a definition of SME that includes firms employing between 1 and 249 people means that considerable variation is likely to exist between SMEs, nevertheless a number of size-related characteristics can be identified:

(i) A limited resource base, particularly with respect to finance and management resources (both management time and a more limited range of management skills) compared with larger firms, because of the more limited scope for managerial division of labour. Indeed, this is one of the arguments that can be used to justify public resources being allocated to developing an external support infrastructure, which small firms can tap into. Limited resources can influence a firm's ability to scan, identify and respond to opportunities and threats presented by the external environment. This includes scanning for new developments relevant to their activity that in a large firm would be typically undertaken by dedicated R&D staff. Limited internal resources is a key justification for interventions designed to provide external support for innovation in SMEs, particularly at the start-up stage or in very young businesses where the opportunities to resource innovation and business development internally are more limited than in established firms.

(ii) A distinctive organizational culture that stems from the combination of ownership and management that typifies the majority of SMEs. This emphasizes the role of the owner-manager and his/her family in the way in which the business is managed and developed, that can affect management behaviour, attitudes to risk, and the nature and extent of external financing. These attitudinal and behavioural characteristics all have potential implications for the nature and extent of the support needs, as well as effective delivery of external support to small companies.

(iii) Less ability to shape and influence the external environment than in the case of larger companies, e.g. relationships with customers, suppliers, sources of finance and the labour market. This means that the smaller firm is typically faced with a more uncertain external environment than a larger firm. As a consequence, competitiveness often relies on the firm's flexibility and adaptability to external changes, which is a key attribute that should be considered and enhanced when designing support programmes and initiatives targeted at SMEs.

The distinctiveness that results from these size-related characteristics affects the 'support needs' of SMEs and the way that support is delivered if it is to be effective. 'Support needs' refers to the need of a firm's management to draw on resources from outside the firm to supply information, advice, training, finance or other assistance which will enable it to deal more effectively with a variety of issues. With respect to innovation, such assistance might range from obtaining basic market information to advice to underpin strategic decisions about new product development that may affect the firm's core business and/or its future development path. In this context, firms may be said to have a 'hierarchy' of support needs which can affect how frequently support is required, how much the firm is prepared to pay for it (if anything), who is the best person/organization to supply it, the firm's ability to make effective use of the support provided, as well as their willingness to accept it. Some of these needs may be provided by private sector organizations as a result of a market transaction; some may be provided by public or semi-public agencies, justified on the basis of a demonstrated market failure in the external support system or because of a recognition by the state of the potential welfare gains to the economy of raising the level and effectiveness of innovative activity; a third possibility is that assistance is delivered by a private sector organization but is partly funded by the state (as in the case of a programme of subsidized consultancy).

One of the issues that arises when considering the 'support needs' of SMEs is the distinction between 'expressed needs' and 'latent needs'. The distinction refers to the difference that can exist between what a business owner or manager's stated wants or expressed needs are compared with what may emerge from a systematic audit of the strengths and weaknesses of the firm in terms of strategies, resources and competencies. Differences between 'expressed' and 'latent' needs often stem from the difficulties which many SME owners/managers have in diagnosing the external support requirements of their business, particularly in cases where they have limited formal professional management training. For example, what may be expressed as a need for marketing support may conceal a deeper latent need for a root-and-branch strategic review of the business, with possible implications for the core business definition. Differences between expressed and latent support needs have important potential implications for policy, particularly where it seeks to be 'client-' or 'market-led'. To some extent, effective demand for business support needs to be developed if the aim of contributing to improving a region's innovative performance and competitiveness is to be achieved.

Although firms of all sizes are likely to require external support of different types from time to time, there are particular characteristics of SMEs that affect both the nature of their external support needs and the process of meeting them effectively. In principle, the more limited resource base of smaller firms with respect to management can make the effective use of external support a necessity if the firm is to be managed effectively. However, the behavioural characteristics of small firms that stem from the combination of ownership and management can result in a reluctance, or even a resistance, to taking in external help for a variety of reasons. These include: doubts about the value for money on the part of the business owner or manager; a scepticism about generalist advice, particularly where this is offered by advisers who lack a detailed sectoral knowledge; and a preference for autonomy which they may perceive is threatened by the use of external advice. This may result in a greater use of informal rather than formal channels of support, in cases where the professional management resources (i.e. where managers lack formal management training or management qualifications) are limited (Smallbone 1997).

It also has implications for the effective delivery of support by formal support agencies because of the importance of trust-based relationships in relation to advice and consultancy in particular. In this regard, previous CEEDR research has drawn attention to the varying impact of different types of external consultancy on SMEs, with assistance with business planning being a more difficult type to deliver effectively than that concerned with more specific issues such as product design/development (Smallbone et al. 1993). The reasons appeared to be a combination of its more fundamental 'root-and-branch' nature and the 'in-and-out' method of delivery typically used by consultants, as well as the more challenging nature of the exercise when previous decisions and judgements of the owner may be placed under the microscope. In such circumstances, the nature of relationship between the external agent or consultant and the client firm is critical to its success. It needs to extend to implementing and monitoring the effectiveness of the assistance offered and not simply based on 'arm's-length' delivery of assistance of a 'advise-and-forget' nature. The distinctiveness of SMEs affects their support needs and how such support is delivered if it is to be effective. In addition, ongoing relationships with external providers of support are more likely to provide a basis for organizational learning than relationships of a one-off sort.

Another important area of support need is finance. No business can develop without adequate and appropriate financial resources and many SMEs face particular difficulties in this respect, as a result of a combination of supply- and demand-side factors that affect their access to both loan and equity finance. Supply-side constraints include an over-emphasis on a collateral-based approach to lending on the part of commercial banks and a reluctance on the part of venture capital fund managers to invest in tranches of less than about £250000 which limits access to these sources for many companies. Whilst a growing recognition of the role of business angels, or informal risk capitalists, exists in some countries, the potential contribution of this source of finance to supporting innovative projects appears to be underdeveloped. It is also important to recognize that SMEs have a range of financial needs with respect to innovation (including working capital as well as investment finance), requiring a mix of finance from different sources, that can present a particular challenge for certain types of company. New or young high-technology-based firms are recognized to face particular problems in this respect (Bank of England 1996). In this context, a key potential role exists for public policy in seeking to address areas where the financial needs of SMEs with respect to innovation are not being adequately addressed through market mechanisms. SMEs also have different financial structures than large firms, typically involving a lower ratio of fixed to total assets. They are more reliant on short-term loans and overdrafts than large firms, where a higher proportion of finance is usually sourced through equity (Cosh and Hughes 1994).

The heterogeneity that exists within the SME sector also has implications for support needs and their effective delivery. One aspect is a firm's sector which leads to some support needs being sector-specific (Curran 1993), although this more commonly applies in manufacturing than in service sector firms (Smallbone et al. 1998). Another important distinction in this regard is between new or very young firms and established firms, since as a firm matures and its management becomes more experienced, its external support needs tend to become more specialized (Smallbone et al. 1993). The problems faced in raising external finance also vary between different stages of business development, with new start-ups facing particular problems because of the absence of a track record and a typically more limited ability to offer sufficient collateral to potential lenders than more established firms.

At the same time, it must be recognized that it is neither realistic nor desirable that all of the support needs of firms (expressed or latent) can or should be addressed through public policy. In this respect, the aim should be to focus those public resources that are available on the needs of the economy at the regional (or national) level. In setting such targets, attention should be paid to the strengths and weaknesses of the innovative capability of the economy, and the priorities for raising it, taking into account the potential welfare gains as well as cost-effectiveness in terms of resource allocation. By implication, there is a need for a regional innovation strategy to guide effective and appropriate decisions in terms of policy development in this area.

1.5   KEY INFLUENCES ON THE INNOVATIVE POTENTIAL OF SMEs

This section focuses on the factors influencing the innovative potential of SMEs and the implications for policy, paying attention to both internal and external influences. However, it is important to stress that the heterogeneity that exists within the SME sector means that there is no single model or set of factors that adequately explains how and why innovation takes place.

The key role played by internal factors on the nature and extent of innovative activity in SMEs has been emphasized by various researchers (see, for example, Hoffman et al. 1998). These internal factors include both personal characteristics of SME owners and managers, such as their background in terms of education and previous experience, and firm characteristics that include both resource and organizational issues, together with the interaction between the two. Both are potentially important to the way that SMEs innovate, not least through their influence on a firm's learning capacity, whether this be internally or through interaction with external individuals and organizations.

In very small and small firms that are owner-managed, an 'organizational learning' perspective often focuses on learning processes in individuals. Drawing upon Personal Construct Theory (Kelly 1955) and contemporary learning theory (Hawkins 1994), Wyer and Boocock (1996) have offered insight into the ways in which small business owner-managers learn and have provided foundations for considering how small firms cope with open-ended change. The thrust of their work is founded on Kelly's proposition that all individuals utilize a personal construct system (derived from inherent personal characteristics and accumulated experience) which is used as a frame of reference to interpret the world. In brief, we all have personal constructs that act as frames of reference to help us view the world which confronts us and to deal with new situations which arise.

If change situations impact on a small firm owner or manager (s)he will use his/her existing personal constructs to cope with the change. On many occasions minor adjustments to the construct may allow the owner-manager to deal with the change, simply because a similar situation has been dealt with in the past. This can be characterized as 'simple' or 'closed loop' learning (Stacey 1996 – see also Argyris and Schon 1978) which takes place in situations where the owner-manager has confirmed the validity of his/her current constructs by using them to make sense of a new situation. The concept may be most relevant to situations where the innovation is incremental rather than novel or radical. However, sometimes change situations arise for which existing constructs are inadequate, which requires them to be extended through a process which entails the questioning of the underlying assumptions upon which the existing constructs are based. This is a more 'complex' or 'double loop' learning process (Stacey 1996), that would appear to be more appropriate to situations where innovation is 'novel' or 'radical' rather than incremental.

Where external knowledge is critical to the innovation process the ability to identify and exploit such knowledge is crucial. Research on cognitive structures and problem-solving has drawn attention to how the learning of individuals is greatest when the new knowledge to be assimilated is related to the individual's existing knowledge structure (Ellis 1965; Estes 1970; Bower and Hildegard 1981). Applying these insights to the organizational level, Cohen and Levinthal (1990) have coined the term 'absorptive capacity' which they define as the firm's general ability to value, assimilate and commercialize new, external knowledge. The term 'receptivity' has been used in a similar way to describe the overall ability of organizations to be aware of, to identify and to take effective advantage of new knowledge (Seaton and Cordey-Hayes 1993).[1] Cohen and Levinthal suggest that an organization's absorptive capacity is largely a function of the firm's prior related knowledge, which in turn is dependent on prior investments in its members' individual absorptive capacities. Investment in R&D is seen as making a particular contribution to a firm's absorptive capacity, although absorptive capacity may also be developed as a by-product of a firm's involvement in other areas, particularly manufacturing. Cohen and Levinthal further argue that the diversity of expertise within a firm is an important source of creativity; as well as strengthening assimilative powers, knowledge diversity facilitates the innovative process by enabling individuals to make novel associations and linkages.

One of the most important determinants of innovative activity at the organizational level relates to the knowledge base of the firm which, in high-technology sectors, may be reflected in a high incidence of qualified scientists and engineers (QSEs) among employees, together with the leadership provided by a highly educated entrepreneur. Management attitudes are another factor since they influence the priority given to innovation in business development as well as in the style of management, particularly with respect to the extent to which there exists a climate in which individual members are encouraged to learn and develop to their full potential. With regard to human capital, an inability to recruit technical staff of sufficient quality can be a serious constraint on innovation for some SMEs, which Cosh et al. (1996) found had significantly increased in importance as a barrier to innovation in SMEs in the UK during the period 1986–95.

Firms also need to be able to commercially exploit the potential benefits of their innovative efforts. Key factors in this respect include the nature and effectiveness of the firm's marketing effort, the degree of marketing involvement in product planning and development and, in some sectors, the firm's competence in the area of technology strategy and technology management (see Hoffman et al. 1998, p. 45; Cobbenhagen et al. 1995). Marketing is a commonly reported weakness in small companies (e.g. Carson 1991) and the marketing of innovative products and services frequently involves developing new geographical markets, including foreign markets, which can represent a particular challenge for smaller firms.

It has long been recognized that most behaviour in organizations is governed by routines (March and Simon 1958; Cyert and March 1963; Nelson and Winter 1982), although it must be emphasized that most of this literature does not explicitly recognize the distinctive organizational characteristics of SMEs. 'Routines' refer to the procedures, conventions, strategies and rules around which organizations are formed and through which they operate. In SMEs, these routines can represent practices developed and nurtured over time to facilitate innovation, or they can reflect the habits and behavioural patterns which evolve in relation to manager's perception of the barriers to innovation. However, unlike many large organizations, these routines are likely to be informal and implicit rather than more formalized and explicitly embedded in management procedures. In other words, in contrast to large firms, most routines in SMEs tend to be tacit and therefore difficult to imitate. In this respect, the capabilities and skills associated with innovation are particularly difficult to emulate. They are often acquired through 'learning by doing' and informal intra-organizational and inter-organizational interactions. In other words, 'learning by interacting' which emphasizes the nature of innovation as a process, is not necessarily perceived in process terms by SME owners and managers.

Management routines, which are effective in supporting innovation, are particularly difficult to acquire, for a number of reasons. Firstly, such routines represent what a firm has learned over time, i.e. what an individual or group has learned over time and handed over to and embedded in the organization and its wider context (Tidd et al. 1997). Secondly, routines are often developed in interaction with other individuals or organizations, such as customers, suppliers and supporting institutions. This applies particularly in the context of more integrated production networks, where innovation and technical change take on a more collaborative and systemic character. As such, routines as well as particular skills of innovation management may be embedded in a specific business network. If the network is broken down, routines will be amputated and cross-skills may be lost, which may explain the long-lasting relations found by Andersen and Christensen (1998) to be predominant between Danish subcontractors and their customers.

The role of networks and long-term relationships is particularly emphasized by writers such as Cooke and Morgan (1998) who suggest that the inclination and ability of SMEs to innovate is linked to the extent to which they enter into interactive learning networks. In this view, the innovative potential of a region depends upon the learning ability of the owners and managers of firms within that region, as well as their technological competence (i.e. their ability to utilize relevant technologies) and entrepreneurial competence (i.e. their ability to integrate relevant technologies with other aspects of the business). Thus, 'interactive learning' and external networking are seen as important stimuli to innovation. In describing innovation as a process of 'know-how' accumulation, Rothwell (1991) says that 'successfully innovative firms generally are well plugged into the market place and to external sources of technological expertise and advice'. A clear link is therefore being drawn between the propensity of SMEs to innovate and their involvement in external networks, although internal factors such as those described above are likely to affect the propensity of firms to engage in such networks.

However, it is not clear, to what extent these 'networks' need to be regional or local as opposed to national or international. A presumption of advocates of the industrial district model of regional economic development is that many traded and untraded interdependencies (e.g. tacit knowledge) tend to operate best at the regional scale. However, some writers have produced evidence that questions this link between innovation and local networking. For example, Hart and Simmie (1997) found that the majority of award-winning innovative firms in Hertfordshire (part of the 'information rich' South East region), did not consider local networks to be important and that a concern for commercial confidentiality prevented firms from co-operating with each other. Moreover, where firms in high-technology sectors have been shown to rely on extensive linkages with a variety of external sources of knowledge, these have been shown to operate over a variety of spatial scales. For example, Keeble et al. (1997) found that technology-intensive firms which have achieved high levels of international links also demonstrate above average levels of local networking with respect to research collaboration and inter-industry linkages.

At the same time, there is evidence which suggests that the role of external sources of knowledge in innovation is complex and varies according to the nature of the activity. Two related areas of investigation are: (a) the contribution of Higher Education Institutions (HEIs) as sources of innovative ideas and technical assistance for SMEs and (b) collaborative research ventures between SMEs or between SMEs and larger firms. There is a body of research which raises a number of concerns about the effectiveness of such linkages in practice. For example, Garnsey and Moore (1993) identify a number of disadvantages for smaller firms with regard to the LINK scheme for supporting collaborative research in the UK, which have wider application. Being more resource-constrained than large firms, small firms often cannot carry the managerial overhead costs required for participation in collaborative research and the scheme is therefore more suited to larger firms. In addition, the independence and adaptability of small firms can be a major source of competitive advantage although these attributes are inevitably constrained when companies have to work in large consortia as required by the LINK scheme. Moreover, under circumstances where it is difficult to protect intellectual property, managers and technologists in small firms express concern about large 'partners' taking over their ideas.

There are a number of policy implications deriving from the insights outlined above. The first is that limited internal resources means that there is an 'a priori' expectation that SMEs might be expected to rely more on external inputs to support innovative activity than larger enterprises. Moreover, the fact that they frequently find it difficult to identify and articulate their own support needs (particularly in terms of technological innovation), highlights the potentially valuable role of intermediaries in offering diagnostic and evaluation support to SMEs (Seaton and Cordey-Hayes 1993;

Oakey and White 1993; Hassink 1996). However, to be successful, these intermediaries need to establish trust-based relations with SME owners/managers. Secondly, small firm owner/managers typically prefer personal and informal links to formal systems for meeting their information and other external support needs, mirroring the less formalized approaches to management within a small organization. Some small firms are receptive to learning from peers with whom they have some shared knowledge base, and therefore a high level of 'relative absorptive capacity' (Lane and Lubatkin 1998). These findings are compatible with those from other sources which suggest that SMEs learn more quickly from other firms; that business partners and customers are the leading stimulants for change and innovation (Rothwell 1991; Dankbaar 1998); and that most SMEs take others as examples, using a reactive strategy, partly imitating others (OECD 1993).

In the case of small owner-managed firms, the transfer of information is most effective between individuals with a similar background (OECD 1993). The third point relates to the valuable role of more resilient channels for developing the absorptive capacity of SMEs and promoting knowledge transfer. In this respect, some of the most effective means of promoting a demand for knowledge, and thus knowledge transfer, involve strengthening the human resource base of the firm, such as by stimulating the employment of graduates in SMEs. Finally, an emphasis on individual and organizational learning has implications for the way that support is delivered to SMEs if it is to be effective. More specifically, it implies a need for innovative interactive approaches to the delivery of training and consultancy to smaller firms in order that the knowledge or skills acquired may lead to a more or less permanent change in behaviour.

It should also be noted that the concept of absorptive capacity (or receptivity) can be used to explain less beneficial effects of inter-organizational learning, such as mimetic isomorphism (DiMaggio and Powell 1983; Lane and Lubatkin 1998) or, applied more broadly to sectors and industrial districts, the phenomena of 'lock-in' (Grabher 1993a). This simply means that while firms learn most easily from peers with whom they have some commonalities this can also lead to patterns of learning which may not be optimal in terms of the firm's longer-term survival, or from a particular regional development perspective which aims to facilitate industrial restructuring by targeting support away from declining sectors towards 'infant industries'. The implications of this will be discussed more fully in Chapter 2.

2. SMEs and the regional dimension of innovation

Bjørn T. Asheim and Arne Isaksen

While Chapter 1 analyses the characteristics of SMEs with respect to innovation, this chapter focuses on another key pillar in this book, which is the role of the regional level in innovation activity. The focus on SMEs and the regional level reflects changes in the approach to innovation policy since the 1980s. One distinctive feature is a greater emphasis on stimulating innovation activity in SMEs, and relatively less focus on the few national 'champions'. Another distinctive feature is a move towards strengthening the regional dimension of policy, and, consequently, the theoretical basis for the increased focus on the regional level is examined in this chapter. However, a broad interpretation of innovation and a more interactive learning conceptualization of the innovation process lie behind both the greater emphasis on SMEs and on the regional level. This new understanding of the innovation process is, thus, further elaborated in this chapter.

2.1 SMEs IN DIFFERENT TYPES OF PRODUCTION AND INNOVATION SYSTEMS

What is meant by the regional dimension of innovation policy? At least two main arguments can be put forward for a strengthening of the regional level in innovation policy. The first refers to the heterogeneity of regions. This is especially recognized in the evaluation of policy instruments in this book, which includes instruments in eleven regions in eight countries, with their different firm and industrial structure as well as distinctive innovation barriers (Chapter 3). With large regional differences there is not one set of policy instruments that suit all types of regions. In order to be effective instruments must be created for or adjusted to differences in regional circumstances. Consequently, one may argue that parts of the national and EU-based innovation policy should be carried out at the regional level where the best knowledge of the varying regional conditions can be secured.

The second argument is based on the view that innovation activity is (also) a territorial phenomenon, meaning that innovation is stimulated by co-operation between local players and by place-specific resources, i.e. resources that are only found in some places, and which cannot rapidly and easily be transferred and 'copied' elsewhere. In these places interactive learning and knowledge spillover take place, resulting in the formation of local relationships, which lead to unique knowledge being created and absorbed in a way that promotes competitiveness for local firms (Storper 1997). This argument about human relationships and specific regional resources underpinning innovation processes is, perhaps, the most important argument for the regionalization of innovation policy. This argument is examined in greater detail in this chapter as it forms an important theoretical basis for the evaluation of innovation policy in the book.

However, as already mentioned in Chapter 1, the significance of the regional level for the innovation activity of SMEs varies among other things according to the different kinds of SMEs. By no means all SMEs are embedded in a local innovative milieu characterized by dense interaction between firms, a high level of collective knowledge etc. SMEs participate in different kinds of production and innovation systems, both regional, national and supra-national. The position and role of firms within such systems will affect the way in which they innovate as well as shape their needs for services from the innovation support system.

Table 2.1 distinguishes between three main types of SMEs and the basic characteristics of their innovation activity. The division builds on the suggestion by Pyke about the existence of 'at least three basic ways in which small firms could seek to survive and prosper in a globalized environment' (Pyke 1994, 4). Small firms could obtain collective strength by networking with other small firms, sometimes firms belonging to the same regional cluster. This implies that firms innovate, produce and/or market through alliances and collective institutions. Secondly, firms could compete more or less 'individually' on final markets. Thirdly, small firms 'could strengthen their claims to be preferred suppliers to large corporations by upgrading their manufacturing quality and delivery standards' (Pyke 1994, 4).

The first main type – SMEs within local production systems – will often make use of local or regional input factors in their innovation process. In some local production systems certain interactive learning takes place, involving local firms and institutions as well as non-local actors, in which unique and 'sticky' knowledge may be created. Parts of this unique knowledge are not 'owned' by any particular firm, but belong to the local production system as a whole as one of its 'intangible' resources (Gottardi 1996). Thus, in local production systems innovations are often the result of firm-specific and region-specific knowledge.

Table 2.1 Characteristics of innovation activity for different types of SMEs

Type of SME	Example	Important source for innovations	Type of innovation system
Firms in local production systems	Industrial districts or other kinds of regional clusters	Local collective knowledge and local actors	Regional, territorially embedded
End firms* outside local production systems	Isolated firms with little or no local collaboration	The R&D-sector for research-intensive SMEs Suppliers of machinery and equipment for less technologically advanced SMEs	National/international
Subcontractors for firms outside the region or for large, local firms	Specialization subcontractors in local innovative milieus Capacity subcontractors in 'low-cost' areas	Local competence and actors Customers	All geographical levels

Notes: * SMEs producing for the final market

Industrial districts provide one example of such local production systems, consisting mainly of SMEs and including (traditionally) both end firms, subcontractors and service firms. The firms enter into formal and informal partnerships, and this co-operation increases the collective innovative capability of these firms. Specialization at local system and firm level as well as co-operation within networks of firms provide the key to innovative activity in industrial districts, making it possible for firms to specialize in core competence and allow their neighbouring firms to carry out complementary activities. This kind of specialization may lead to high levels of competence amongst groups of firms, within relatively narrow fields, which in turn increases the chances of identifying new, cost-efficient solutions. Employees can, for example, discover better production methods, or identify new product solutions. Innovative capacity is further strengthened by the establishment of a regional research and education system directed towards the dominant branches of the region.

It is the first kind of SMEs in Table 2.1 that most typically rely on resources of the regional industrial milieu for their competitiveness – and the basis for this kind of competitiveness is examined in more detail in the rest of the chapter. However, we shall first discuss the two other types of SMEs that are less embedded or integrated in regional production and innovation systems. Thus, another main type of SMEs is the 'isolated' end firm, i.e. firms that do not – or cannot – participate in local production systems. Such firms are sometimes found in peripheral areas where there are few suitable local firms to co-operate with, but the firms may enter into collaboration in national and/or international production and innovation systems.

Research-intensive, 'isolated' SMEs in particular, may be integrated in innovation systems at higher geographical levels. These are resource-strong firms that are keen to co-operate with R&D organizations, or which can enter into strategic alliances with advanced firms outside their home region to promote economic growth. By way of education and job experience the entrepreneurs, firm managers and/or engineers may be members of a community of professionals, which facilitates the interchange of information and knowledge between persons independent of location. Research-intensive or high-tech small firms are typically spin-offs from universities, R&D institutes or other small or large high-tech firms, and the spin-offs are often located near their 'mother' organization. Thus, research-intensive firms tend to have dense interactions with neighbouring knowledge organizations and firms. When these kinds of firms are located outside a regional innovative milieu, they may thus be integrated in a 'professional' space, i.e. interacting with players belonging to some other regional innovation systems, or being part of national and international systems.

Other 'isolated' firms may be less integrated in national and international innovation systems than the research-intensive ones. Thus, incrementally innovative and non-innovative SMEs often lack the necessary competence to co-operate directly with R&D organizations and R&D-intensive firms. Such SMEs may not have highly educated employees who 'speak the same language' as researchers at R&D institutes, and they may lack both the competence and capital necessary to carry out R&D projects on their own. However, R&D knowledge may play an important role in the innovation processes of these firms, although in an indirect way, for example via ubiquitous knowledge drawn from a national or global knowledge base (Smith 1999). These knowledge bases are accessed via national and international suppliers of equipment, with their related consulting activities including installation, test running, service and maintenance, training and skill development. Contact and co-operation with suppliers and traders of production equipment and inputs is likely to be an important source for innovation support in 'isolated' and less resource-rich SMEs, in addition to interaction with customers and clients and the firm-specific knowledge built up within the firms.

The third SME-type in Table 2.1 refers to subcontractors who supply firms outside the region, or large, dominant local firms. In principle, these SMEs can belong to one of two main types, defined according to the relationship between subcontractor and customer. The first type of subcontractors co-operates with customers on design and quality, and often employs long-term contracts (Amin and Robins 1991). Such subcontractors generally have their own product-range, are highly technically competent and are in a strong position in relation to their customers (Grabher 1993a). Known as specialization subcontractors (Holmes 1986), they hold additional and complementary core competence with respect to their customer firms – competence their customers often need for their own innovation activity.

Specialization subcontractors can participate in many different kinds of innovation systems. They may enter into co-operation with national or global innovation systems in those cases where they have clients outside the local region. Specialization subcontractors must be innovative to survive; they are under pressure from customers to constantly upgrade their products' technological base, and also to take up new production concepts such as 'just-in-time' deliveries. In some cases these suppliers may have attained high levels of technical competence and innovative ability through initially being located within innovative, local production systems. Thus, transnational corporations (TNCs) to a certain extent identify their suppliers in different knowledge-intensive milieus according to their need to connect their own knowledge bases with locally based, often tacit and immobile competence rooted in innovative regional industrial milieus (Mariussen 1997). TNCs are increasingly aware of the potential of exploiting unique, local knowledge and creativity in specific places as a source of profit (Davis 1995). The pursuit of this kind of strategy by TNCs extends the reach of global production systems by simultaneously linking global and regional innovation systems. Thus, firms become competitive through the mobilization of location-specific resources in different places, for example by tapping into other firms' expertise through subcontracting, take-overs or strategic alliances (Storper 1997). Furthermore, competence and technology from global production systems may also diffuse to other firms in local networks.

Dependent subcontractors are in a very different position compared to specialization subcontractors. The former have very little technical competence, produce components only to order, are subject to strong pressure on pricing and are in constant danger of being rejected in favour of other subcontractors (Grabher 1993a). What we see here is an asymmetric relationship between the customers and the capacity subcontractors. This kind of subcontractor competes through lowering prices, and exploiting numerical flexible working arrangements such as short-term contracts, overtime, putting out and subcontracting to other lower-tier firms. These firms are often incrementally innovative or non-innovative firms, and, thus, not part of a regional innovation system.

2.2   INTERACTIVE LEARNING, CLUSTERS AND REGIONAL INNOVATION SYSTEMS

The chapter now goes on to describe in more detail the territorial dimension of innovation, mostly related to SME-type number one in Table 2.1. The important starting point is, thus, to understand both industrialization as a territorial process, i.e. underlining the importance of agglomeration and 'non-economic' factors for economic development, and innovation as a socially embedded process, i.e. as an institutionally and culturally contextualized, interactive learning process. The combined effect of these two processes has changed the view on SMEs as a strategic job-generating instrument in future regional and industrial policy. While innovative, non-high-tech SMEs previously were looked upon as dynamic but fragile exceptions from the modern high-tech based path of industrialization, this new understanding looks at innovative and competitive SMEs as a result of successful regionalization strategies, i.e. as an alternative way of achieving global competitive advantage. This regionalization strategy is based on:

(i) learning as a localized process, pointing at the importance of histori-
     cal trajectories and 'disembodied knowledge';
(ii) innovation as an interactive learning process, involving a critique of
     the linear model of innovation and emphasizing the importance of
     co-operation in promoting competitiveness; and
(iii) agglomeration as the most efficient basis for interactive learning,
     arguing for the importance of 'untraded interdependencies' and
     bottom-up, interactive regional innovation systems and networks.

In much of the literature on industrial districts during the last 10–15
years the seemingly paradoxical productive role played by traditional, pre-
capitalistic socio-cultural structures in competitive, modern local and
regional economies has been discussed as well as questioned. Commenta-
tors generally agreed that what made these regions (e.g. industrial districts)
so successful was their combination of functional and territorial integration. The territorial dimension of the socio-cultural structures represented
the basic input promoting flexibility and dynamism. However, on the one
hand, the continual influence of socio-cultural structures was said to make
the regions vulnerable to changes in the global capitalist economy, but on
the other hand much work was put into the evaluation of the adaptability
and replicability of the industrial district model to other regions in need of
development strategies (Asheim 1994).

An important factor contributing to the generalization of the experiences of industrial districts is the new theoretical understanding of innovation as basically a social process. Compared to the previous dominating
linear model of innovation, this implies a more sociological view, in which
interactive learning is looked upon as a fundamental aspect of the innovation process, which thus cannot be understood independent of its institutional and cultural contexts (Lundvall 1992).

Taken together, this theoretical development has dramatically changed
the basis for launching innovation policies towards SMEs with the intention of promoting endogenous regional development. Such a regionalization approach can be seen as an alternative strategy of achieving
competitiveness in a global economy, a position which has often been
neglected in the globalization debate.

2.2.1 Towards an Interactive Innovation Model

One important type of innovation policy is the formation of innovation
systems. Basically, an innovation system consists of a production structure
and an institutional infrastructure, and the interaction between these structures. Innovation systems are normally referred to as national systems, but

they can also comprise larger or smaller geographical areas. In recent years we have seen an increasing interest in regional innovation systems, in particular, both by academics and policy-makers (Storper 1995).

The increased interest in regional innovation systems is also a result of the new theoretical understanding of the innovation process, which points to new options for SMEs in innovation performance, and new possibilities in designing innovation policy aiming at SMEs. Traditionally, policies for upgrading the innovative capability of SMEs have been based on introducing (more) formal R&D-based product and process innovations. The problem with this strategy, however, has been that formal R&D activity has normally been out of reach for the majority of SMEs due to lack of financial as well as human resources (Chapter 1). Traditional SMEs have a more limited resource base (particularly finance and management) compared to larger firms. However, some SMEs are very innovation-rich, especially SMEs in high-tech sectors, employing many persons with higher education and having extensive co-operation with R&D institutions. At the same time the key potential competitive strength, in general, stems from SMEs' adaptability and flexibility, which tends to point to a non-linear model of support.

Modern innovation theory has developed as a result of criticism of the traditional dominating linear model of innovation, as the main strategy for national R&D policies, of being too 'research-based, sequential and technocratic' (Smith 1994, 2). This criticism implies another and broader view of innovation as a social as well as a technical process, as a non-linear process, and as a process of interactive learning between firms and their environment (Lundvall 1992, Smith 1994). In fact it could be argued that due to the rapid technological change characterizing the globalizing learning economy, the linear model, which is timely and costly, can only be used efficiently in basic research in laboratories of universities and large firms in such R&D-intensive branches as pharmaceuticals and defence industry. Thus, this cannot, in general, be seen independently of the type of industry in question, as high-tech industries to a larger degree will continue to be most dependent on formal R&D. As a consequence, due to its basic characteristics, it will remain expensive and protected, and, will not promote co-operation. However, in industries with expensive, but not especially advanced R&D-based product innovations, such as in the automobile industry, more co-operation has been applied in order to share development costs.

This alternative model could be referred to as a bottom-up interactive innovation model (Asheim and Isaksen 1997), much more adapted to traditional SMEs and the 'learning economy', where knowledge is the most fundamental resource and learning the most important process (Lundvall and Johnson 1994). Lundvall and Johnson use the concept of 'learning economy' when referring to the contemporary post-Fordist economy dominated by the ICT (information, computer and telecommunication) -related techno-economic paradigm (Lundvall and Johnson 1994). In addition to the combined effect of widespread ICT technologies and flexible production methods, the learning economy is firmly based on 'innovation . . . (understood as interactive learning) . . . as a crucial means of competition' (Lundvall and Johnson 1994, 26). The interactive innovation model puts emphasis on 'the plurality of types of production systems and of innovation (science and engineering is only relevant to some sectors), "small" processes of economic co-ordination, informal practices as well as formal institutions, and incremental as well as large-scale innovation and adjustment' (Storper and Scott 1995, 519).

What this broader understanding of innovation as a social, non-linear and interactive learning process means, is a change in the evaluation of the importance and role played by socio-cultural structures in regional development from being looked upon as mere reminiscences from pre-capitalist civil societies (although still productive), to be viewed as necessary prerequisites for regions in order to be innovative and competitive in a post-Fordist learning economy. According to Amin and Thrift, this forces a re-evaluation of 'the significance of territoriality in economic globalization' (Amin and Thrift 1995, 8). Furthermore, this new and alternative conceptualization of innovation as an interactive learning process means an extension of the range of branches, firm sizes and regions that can be viewed as innovative, also to include traditional, non R&D-intensive branches, often constituted by SMEs and located in peripheral regions. The basic critique of the linear model is precisely the equation of innovative activities with R&D intensity. The majority of SMEs are in branches which are not R&D-intensive, but which could still be considered to be innovative (e.g. the importance of design in making furniture manufactures competitive and moving them up the value-added chain). Table 2.2 provides a summary of characteristics of the two innovation models.

The emphasis on interactive learning as a fundamental aspect of the process of innovation points to co-operation as an important strategy in order to promote innovations (Asheim 1996). In interactive innovation processes interactive learning takes place (a) between different steps of the innovation process, involving the mobilization of different forms of knowledge and information (e.g. science-based knowledge, market information, technical skills); (b) with different firms and organizations involving inter-firm collaborations between suppliers and subcontractors as well as with customers; (c) with different knowledge production centres and organizations, representing a wide variety from R&D institutions via other parts of the knowledge infrastructure broadly defined to other firms or departments within a corporation (i.e. if the firm belongs to a TNC); and (d) interaction between different departments of the same enterprise, involving the co-operation between different groups of employees with different forms of knowledge (e.g. R&D-based, artisan and tacit knowledge) (Asheim 1999b; Lundvall and Borrás 1997).

Table 2.2   Characteristics of two innovation models

	Linear innovation model	Interactive innovation model
Important players	Large firms and the R&D sector	Both small and large firms, the R&D sector, clients, suppliers, technical colleges, public authorities
Important inputs in the innovation process	R&D	R&D, market information, technical competence, informal practical knowledge
Geographical consequences	Most innovative activity (R&D) in central areas	Innovation activity more geographically widespread, but especially occurring in (different types of) regional clusters
Typical industrial sectors	Fordist manufacturing	Flexible industrial sectors
Implications for regional policy	Promote R&D in less central areas Promotion of technological diffusion	Develop regional innovation systems, and link firms to wider innovation systems

Source: Asheim and Isaksen 1997.

The hegemonic techno-economic paradigm of the post-Fordist 'learning economy' is to a very large extent (and more than previous techno-economic paradigms) dependent on organizational innovations to have its potential exploited. The more important organizational innovations are, the more important interactive learning can be considered to be for the promotion of innovations in general, as organizational innovations enable the formation of learning organizations in post-Fordist societies. A dynamic flexible 'learning organization' can be defined as one that promotes the learning of all its members and has the capacity of continuously transforming itself by rapidly adapting to changing environments by adopting and developing innovations (Pedler et al. 1991, Weinstein 1992). Such learning organizations must ideally be based on strong involvement of workers within firms, on horizontal co-operation between firms in networks, and on bottom-up, interactive-based innovation systems at the regional level and beyond. This could, together with other necessary organizational and institutional innovations at different administrative levels from the local to the supra-national, contribute to the formation of 'learning regions' (Asheim 1996, 2001).

Firms of the learning economy are basically 'learning organizations'. They choose organizational modes such as inter-firm networking and intra-firm horizontal communication patterns in order to enhance learning capabilites (Lundvall and Johnson 1994). Lundvall and Johnson argue that 'the firm's capability to learn reflects the way it is organized. The movement away from tall hierarchies with vertical flows of information towards more flat organizations with horizontal flows of information is one aspect of the learning economy' (Lundvall and Johnson 1994, 39). This is in line with Scandinavian experiences, based on the socio-technical approach to organization theory, which have shown that flat and egalitarian organizations have the best prerequisites of being flexible and learning organizations, and that industrial relations characterized by strong involvement of functional flexible, central workers is important in order to have a working 'learning organisation' (Asheim 1996, 2001).

Thus, if these observations are correct, this represents new 'forces' in the promotion of technological development in capitalist economies, implying a modification of the overall importance of competition between individual capitals. Of course, the fundamental forces in a capitalist mode of production constituting the technological dynamism are still caused by the contradictions of the capital-capital relationship. However, Lazonick argues, referring to Porter's empirical evidence (Porter 1990), that 'domestic co-operation rather than domestic competition is the key determinant of global competitive advantage. For a domestic industry to attain and sustain global competitive advantage requires continuous innovation, which in turn requires domestic co-operation' (Lazonick 1993, 4). Cooke (1994) supports this view, emphasizing that 'the co-operative approach is not infrequently the only solution to intractable problems posed by globalization, lean production or flexibilization' (Cooke 1994, 32).

2.2.2 Knowledge Infrastructures and Interactive Learning

One further, important implication of this view on innovation is that it makes the distinction between high-tech and low-tech branches and sectors, which is a product of the linear model, irrelevant when competitiveness is discussed, as it maintains that all branches and sectors can be innovative in this broader sense, although they innovate differently. According to Porter, 'the term high-tech, normally used to refer to fields such as information technology and biotechnology, has distorted thinking about competition, creating the misconception that only a handful of businesses compete in sophisticated ways. In fact, there is no such thing as a low-tech industry. There are only low-tech companies – that is, companies that fail to use world-class technology and practices to enhance productivity and innovation' (Porter 1998, 85–6). Following Porter, this implies that it is possible in all branches and sectors to find productive and innovative firms enjoying competitive advantages on the global markets. Thus, this theoretical perspective even broadens the scope for a policy of strong competition for post-Fordist learning economies (Storper and Walker 1989), i.e. competition building on innovation and differentiation strategies, in contrast to weak competition based on price competition.

Thus, instead of using high-tech and low-tech to describe firms and branches, it would be more theoretically adequate and empirically relevant to talk about the distributed knowledge base of firms and the knowledge infrastructure of branches and regions, in order to better understand the complex interactions and relationships which characterize the innovation processes of firms in the vertical disintegrated, global and local production systems of the post-Fordist learning economy. This points to the importance of the knowledge infrastructures of regions and countries. According to Smith, 'any analysis of the technological performance of a country or region should therefore have the infrastructure clearly in focus' (Smith 1997, 94). Knowledge infrastructures are constituted by a variety of institutions and organizations such as universities, other R&D institutions, training systems, production knowledge of firms etc. 'whose role is the production, maintenance, distribution, management, and protection of knowledge' (Smith 1997, 94–5).

Such knowledge infrastructures are of strategic economic importance concerning the promotion of innovation and economic growth, since all industrial production is based on knowledge, which can be either formal, codified (scientific or engineering knowledge) or informal, tacit (embodied in skilled personal routines or technical practice) (Smith 1997). Of specific importance in the context of localized learning is what could be called 'soft' knowledge infrastructures, i.e. infrastructures producing knowledge according to an interactive, bottom-up model – for regional economic performance.

One of the consequences of the considerably more knowledge-intensive modern economies is that 'the production and use of knowledge is at the core of value-added activities, and innovation is at the core of firms' and nations' strategies for growth' (Archibugi and Michie 1995, 1). Thus, in a 'learning economy' 'technical and organizational change have become increasingly endogenous. Learning processes have been institutionalized and feed-back loops for knowledge accumulation have been built in so that the economy as a whole [. . .] is "learning by doing" and "learning by using"' (Lundvall and Johnson 1994, 26). Lundvall and Borrás explicitly argue that they prefer 'the learning economy' to 'the knowledge-based economy' as it 'emphasizes the high rate of economic, social and technical change that continuously underlies specialized (and codified) knowledge. It makes it clear that what really matters for economic performance is the ability to learn (and forget) and not the stock of knowledge' (Lundvall and Borrás 1997, 35). However, as knowledge, according to Lundvall and Johnson (1994), is considered the most fundamental resource, the learning economy is of course a knowledge-based economy. Furthermore, in order to underline the dynamic and rapid change in the contemporary globalizing economy it is necessary also to pay attention to knowledge creation as a process of equal importance to learning and forgetting. Nonaka and Reinmöller emphasize that 'organizational and interorganizational analysis of the conditions for innovation underlines the importance of knowledge and the key process of knowledge creation' (Nonaka and Reinmöller 1998, 410).

One problematic aspect of the 'learning economy' has been its focus being mainly on 'catching up' learning (i.e. learning by doing and using) based on incremental innovations, and not on radical innovations requiring the creation of new knowledge. In a long-term perspective of the globalizing economy it will be increasingly difficult for the reproduction and growth of a learning economy to primarily rely on incremental improvements of products and processes, for example in the form of imitation, and not on basically new products (i.e. radical innovations) as a result of, for example, an invention, even if Freeman underlines 'the tremendous importance of incremental innovation, learning by doing, by using and by interacting in the process of technical change and diffusion of innovations' (Freeman 1993, 9–10). Focusing on territorial agglomerated SMEs, Crevoisier argues that the reliance on incremental innovations 'would mean that these areas will very quickly exhaust the technical paradigm on which they are founded' (Crevoisier 1994, 259), and Bellandi sees 'the assessment of the endogenous innovation capacities of the industrial districts . . . [as] . . . a key issue' (Bellandi 1994, 93). More specifically, this means the capability to break path dependence and change technological trajectory through radical innovations, so as to avoid falling into 'lock-in situations' as a result of internal 'weakness of strong ties' (Granovetter 1973) or of external 'weak competition' from low-cost producers (Glasmeier 1994). This would be even more important if the alternative strategy was to base regional development on exogenous learning. According to Nonaka and Reinmöller, 'no matter how great the efficiency and speed of exogenous learning, it will not substitute for the endogenous creation of knowledge.

The faster knowledge is absorbed, the greater the dependence on the sources of knowledge becomes' (Nonaka and Reinmöller 1998, 425–6). Thus, what is more and more needed in a competitive globalizing economy is the creation of new knowledge through searching, exploring and experimentation involving creativity as well as more systematic R&D in the development of new products and processes.

2.2.3 Learning as a Localized Process

In the perspective of this new understanding of innovation, strategic parts of learning processes emerge as a localized, and not as a placeless, process, and, thus, constitute important parts of the knowledge base and infrastructure of firms and regions, which points to the role of historical trajectories. This view is supported by Porter, who argues that 'competitive advantage is created and sustained through a highly localized process. Differences in national economic structures, values, cultures, institutions, and histories contribute profoundly to competitive success' (Porter 1990, 19). Accordingly, Porter argues that 'the building of a "home base" within a nation, or within a region of a nation, represents the organizational foundation for global competitive advantage' (as referred in Lazonick 1993, 2). Localized learning is not only based on tacit knowledge, as we argue that contextual knowledge also is constituted by 'sticky', codified knowledge. This refers to 'disembodied' knowledge and know-how which are not embodied in machinery, but are the result of positive externalities of the innovation process, and generally based on a high level of individual skill and experience, collective technical culture and a well-developed institutional framework, which are highly immobile in geographical terms (de Castro and Jensen-Butler 1993), and, thus, can represent important context conditions of regional clusters with a potentially favourable impact on their innovativeness and competitiveness. Such 'disembodied' knowledge is often constituted by a combination of place-specific experience-based, tacit knowledge and competence, artisan skills and R&D-based knowledge (Asheim 1999a).

Disembodied knowledge can, thus, be both tacit and codified, which implies that some codified knowledge can be a product of localized rather than placeless learning. This implies that the adaptability of this localized form of codified knowledge is dependent upon, and limited by, artisan skills and tacit knowlege (Asheim and Cooke 1998). In a similar way, Malmberg (1997) argues that 'one of the few remaining genuinely localized phenomena in this increasingly "slippery" global space economy is precisely the "stickiness" of some forms of knowledge and learning processes' (Malmberg 1997, 574; Markusen 1996), and Lundvall maintains that 'the increasing emergence of knowledge-based networks of firms, research groups and experts may be regarded as an expression of the growing importance of knowledge which is codified in local rather than universal codes' (Lundvall 1996, 10).

Other researchers have also recognized the need for an intermediate form of contextual knowledge transcending the dichotomy of codified and tacit knowledge. Nonaka and Reinmöller maintain that 'industrial regions can provide the necessary combination of explicit knowledge and tacit knowledge through colocation' (Nonaka and Reinmöller 1998, 421), and Lundvall and Borrás argue that 'tacit knowledge may be shared through human interaction and this may be the major force behind the formation of business networks. This means that codified and tacit knowledge are complementary and co-exist in time' (Lundvall and Borrás 1997, 33).

Following this line of reasoning it could be argued that the combination of contextual disembodied knowledge and 'untraded interdependencies', i.e. 'a structured set of technological externalities which can be a collective asset of groups of firms/industries within countries/regions' and which represent country- or region-specific 'context conditions' of fundamental importance to the innovative process (Dosi 1988, 226), can constitute the material basis for the competitive advantage of regions in the globalizing learning economy. Storper (1997) defines such contexts as 'territorialization', understood as a distinctive subset of territorial agglomerations, where 'economic viability is rooted in assets (including practices and relations) that are not available in many other places and cannot easily or rapidly be created or imitated in places that lack them' (Storper 1997, 170). This would represent an argument against the idea that 'ubiquitification' (i.e. the global availability of new production technologies and organizational designs at more or less the same cost (Malmberg and Maskell 1999) as an outcome of globalization and codification processes), in general tends to 'undermine the competitiveness of firms in the high-cost areas of the world' (Malmberg and Maskell 1999, 6). Such an argument is implicitly based on the dominance of a near free-market situation in the global economy, leaving no room for the importance of networks and clusters, creating external economies and increasing returns, as the economic basis for imperfect competition (Krugman 1991), as well as on the principles of comparative advantage, based on cost advantages, for example, through the exploitation of a supply of cheap labour (Porter 1998).

Concerning the question of the extent of codification of tacit knowledge in the globalizing learning economy, Lundvall and Borrás argue that 'there are two important limits to the codification process. First, the fact that codified and tacit knowledge are complementary and co-existing means that there are natural limits to codified knowledge. . . . And second, increased codification does not necessarily reduce the relative importance of tacit knowledge – mostly skills and capabilities – in the process of learning and knowledge accumulation. Actually, easier and less expensive access to information makes skills and capabilities relating to the selection and efficient use of information even more crucial than before. This means that tacit knowledge is still a key element in the appropriation and effective use of knowledge, especially when the whole innovation process is accelerating' (Lundvall and Borrás 1997, 33).

Thus, the strict dichotomy normally applied between codified and tacit knowledge can be quite misleading both from a theoretical as well as from a policy point of view. This is especially the case if localized learning is primarily said to be based on tacit knowledge. Thus, we agree with Lundvall and Borrás, who claim that 'it is the constitution of new ensembles of codified and tacit knowledge which is in question rather than a massive transformation of tacit into codified knowledge' (Lundvall and Borrás 1997, 33).

2.2.4   Clusters, Networks and the Competitive Advantage of Regions

A dynamic, processual understanding of competitiveness clearly indicates that enterprises in order to keep their position in the global market must focus on developing their own core competencies (which also includes new competencies) through transforming themselves into learning organizations. But internal restructuring alone cannot sustain the competitiveness of firms in the long run. As firms are embedded in regional economies (although in a varying degree) they are very much dependent on a favourable economic and industrial environment in general, and knowledge infrastructures at different geographical levels specifically. According to Porter, untangling the paradox of location in a global economy reveals a number of key insights about how companies continually create competitive advantage. What happens inside companies is important, but clusters reveal that the immediate business environment outside companies plays a vital role as well (Porter 1998, 78).

Thus, a strong case is made today in favour of regional clusters growing in importance as a mode of economic co-ordination in post-Fordist learning economies (Asheim and Isaksen 1997, Cooke 1994). The main argument for this is that regional clusters may provide an optimal context for an innovation-based learning economy due to the existence of localized learning and 'untraded interdependencies' among actors. In general, 'geographical distance, accessibility, agglomeration and the presence of externalities provide a powerful influence on knowledge flows, learning and innovation and this interaction is often played out within a regional arena' (Howells 1996, 18). Close co-operation with suppliers, subcontractors, customers and support institutions in the region, based on human relationships, may enhance the process of interactive learning and create an innovative milieu favourable to innovation and constant improvement. This influences the performance of the firms and strengthens the competitiveness of the clusters, and is increasingly seen as an important aspect of fostering regional competitive advantage.

Agglomeration economies can thus represent important basic conditions and stimuli to incremental innovations through informal 'learning-by-doing' and 'learning-by-using', primarily based on tacit knowledge (Asheim 1994). As Bellandi suggests, such learning, based on practical knowledge (experience) of which specialized practice is a prerequisite, may have significant creative content, implying that the collective potential innovative capacity of small firms in industrial districts is not always inferior to that of large, research-based companies (Bellandi 1994). Still the fact remains that, in general, the individual results of what he calls decentralized industrial creativity (DIC) are incremental, even if 'their accumulation has possible major effects on economic performance' (Bellandi 1994, 76).

This perspective on the importance of regional clusters can find support from modern innovation theory, originating from new institutional economics, which argues that 'regional production systems, industrial districts and technological districts are becoming increasingly important' (Lundvall 1992, 3), and from Porter, who emphasizes that 'the process of clustering, and the interchange among industries in the cluster, also works best when the industries involved are geographically concentrated' (Porter 1990, 157). In 1998 Porter argued strongly that 'a vibrant cluster can help any company in any industry compete in the most sophisticated ways, using the most advanced, relevant skills and technologies' (Porter 1998, 86).

Thus, what is a cluster? In an article in 1998 Porter defines clusters as 'geographic concentrations of interconnected companies and institutions in a particular field. Clusters encompass an array of linked industries and other entities important to competition. They include, for example, suppliers of specialized inputs such as components, machinery, and services, and providers of specialized infrastructure. Clusters also often extend downstream to channels and customers and laterally to manufacturers of complementary products and to companies in industries related by skills, technologies, or common inputs. Finally, many clusters include governmental and other institutions – such as universities, standards-setting agencies, think tanks, vocational training providers, and trade associations – that provide specialized training, education, information, research, and technical support' (Porter 1998, 78).

As a contrast, Porter's original cluster concept was basically an economic concept indicating that 'a nation's successful industries are usually linked through vertical (buyer/supplier) or horizontal (common customers, technology etc.) relationships' (Porter 1990, 149). In our view there is a need to operate with clusters in both conceptualizations, as it is a quite normal situation to find (geographical) clusters of specialized branches being part of a national (economic) cluster of the same branches (e.g. the Norwegian shipping cluster, which is a national economic cluster (Reve et al. 1992), but which, in part, is constituted by geographical clusters of specialized branches making up the Norwegian shipping cluster).

What this extension of the definition of the concept of cluster also indicates is a deepening and widening of the degree and form of co-operation taking place in a cluster. The original and simplest form of co-operation within a cluster can often be described as territorial integrated input–output (supply chain) relations, which could be supported by informal, social networking as is the case with Marshallian agglomeration economies, but which could also take the form of arm's-length market transactions between a capacity subcontractor and the client firm. The next step of formally establishing inter-firm networks is represented by a purposeful, functional integration of value chain collaboration in production systems as well as building up a competence network between the collaborating firms, which could form part of a regional innovation system. A distinction between clusters defined as input–output relations and networks is that proximity is the most important constituting variable in the first case, while formal networking represents a step towards more systemic (i.e. planned) forms of co-operation, as well as a development from vertical to horizontal forms of co-operation, which more efficiently promotes learning and innovation in the systems. Or as Nonaka and Reinmöller put it, 'industrial districts are accumulations of interdependent companies located near each other (the condition of proximity). Networks are a concept focused on inter-organizational relations. . . . Unlike the concept of industrial districts, the concept of networks does not necessarily entail the condition of proximity' (Nonaka and Reinmöller 1998, 406).

The new, post-Fordist ways of organizing industrial production can take various forms. The specific new form of industrial organization resulting from close inter-firm networking is represented by 'quasi-integration' (Leborgne and Lipietz 1988). Quasi-integration refers to relatively stable relationships between firms, where the principal firms (i.e. the buyers) aim at combining the benefits of vertical integration as well as vertical disintegration in their collaboration with suppliers and subcontractors (Haraldsen 1995). According to Leborgne and Lipietz 'quasi-integration minimizes both the costs of co-ordination (because of the autonomy of the specialized firms or plant), and the costs of information/transaction (because of the routinized just-in-time transactions between firms). Moreover the financial risks of R&D and investments are shared within the quasi-integrated network' (Leborgne and Lipietz 1992, 341).

Leborgne and Lipietz (1992) maintain that the more horizontal the ties between the partners in the network are, the more efficient the network as a whole is. This is also emphasized by Håkansson, who points out that 'collaboration with customers leads in the first instance to the step-by-step kind of changes (i.e. incremental innovations), while collaboration with partners in the horizontal dimension is more likely to lead to leap-wise changes (i.e. radical innovations)' (Håkansson 1992, 41). Generally Leborgne and Lipietz argue that 'the upgrading of the partner increases the efficiency of the whole network' (Leborgne and Lipietz 1992b, 399).

This reorganization of networking between firms can be described as a change from a domination of vertical relations between principal firms and their subcontractors to horizontal relations between principal firms and suppliers. Patchell refers to this as a transformation from production systems to learning systems, which implies a transition from 'a conventional understanding of production systems as fixed flows of goods and services to dynamic systems based on learning' (Patchell 1993, 797). Such institutionalization of a continual organizational learning process involves a redefinition of a firm's relations to its major suppliers based on the recognition that 'a network based on long-term, trust-based alliances could not only provide flexibility, but also a framework for joint learning and technological and managerial innovation. To be an integral partner in the development of the total product, the supplier must operate in a state of constant learning, and this process is greatly accelerated if carried out in an organizational environment that promotes it' (Bonaccorsi and Lipparini 1994, 144). According to Lundvall 'the growing complexity of the knowledge base and the more rapid rate of change makes it attractive to establish long-term and selective relationships in the production and distribution of knowledge. The skills necessary to understand and use these codes will often be developed by those allowed to join the network and to take part in a process of interactive learning. Perhaps one of the most fundamental characteristics of the present phase of the learning economy is the formation of knowledge-based networks some of which are local while others cross national boundaries' (Lundvall 1996, 10–11).

Such enterprise development is based on what Lazonick and O'Sullivan (1996) call the innovative enterprise. An innovative business enterprise does not take an achieved competitive advantage as given, as it can be eliminated through imitation. Thus, it must be continually reproduced through innovation. However, the innovation process has to be based on collective learning inside the business enterprise or network of co-operating firms to give the firm a possibility of developing their specific competitive advantage over competing enterprises. In this way collective learning stands in contrast to individual learning, where the improved skills are sold and purchased on the labour market at a given price (Lazonick and O'Sullivan 1996; Storper 1997). According to Lazonick and O'Sullivan (1996), innovation processes in the advanced knowledge-based society are characterized by such collective learning, which depends on business enterprises creating social organizations (e.g. learning organizations and networks) enabling collective learning to take place.

2.2.5   Types of Regional Innovation Systems and Barriers

The growing interest in the role of national and regional innovation systems must be understood in the context of creating a policy instrument aiming at a systematic promotion of localized learning processes in order to secure the innovativeness and competitive advantage of national and regional economies (Freeman 1995; Cooke 1995). According to Storper and Scott, 'a new "heterodox" economic policy framework has emerged in which significant dimensions of economic policy at large are being reformulated in terms of regional policies' (Storper and Scott 1995, 513). This is partly the result of the economic success stories of territorially agglomerated clusters of SMEs (e.g. in the Third Italy), which have become a major point of reference in the recent international debate on industrial policy promoting endogenous development, and partly the result of the new political initiatives towards a 'Europe of regions', where the development prospects of the lagging regions of Europe in particular have been of great concern to the EU.

The concept of innovation system is based on the idea that the overall innovation performance of an economy to a large extent depends on how firms manage to utilize the experience and knowledge of other firms, research organizations, the government sector agencies etc. in innovation processes, and not just on the capability of the individual firm (although the competencies and attitude of entrepreneurs, managers and workers are also important for their innovation capability). Thus, factors stimulating innovativeness to a considerable degree seem to be determined by conditions in the environment of firms, and specific contextual factors may hamper as well as promote innovation processes. However, the environment of firms should be understood both in a territorial and functional sense. In a functional sense firms draw on ideas, know-how and complementary assets from customers, suppliers, consultants, universities, funding and training organizations, independent of geographical location (Tödtling and Kaufmann 1999). SMEs are also often linked up to the global economy through international production and innovation systems, where other firms, universities and R&D institutes can represent accessibility to global knowledge. In particular, regional clusters of SMEs need to be 'in touch, not necessarily directly, but through the supply chain with global networks' (Cooke 1998, 10) in order to attract the complementary assets needed to be competitive, when, for example, local R&D competence may be scarce.

In a territorial sense, the stock of knowledge and the learning ability in the regional industrial milieu can be of great importance in stimulating the innovation capability of firms. SMEs in particular seem to depend on assets in the regional industrial milieu when innovating, as by definition they often have scarce resources internally, as well as external problems in managing collaboration with remote actors. Thus, 'smaller firms – particularly those that lack resources and incentives to develop their own training, research or engineering departments – depend heavily on local services' (Rosenfeld 1997, 20). As underlined throughout the chapter, regions are more generally seen as an important unit of economic co-ordination at the meso level: 'the region is increasingly the level at which innovation is produced through regional networks of innovators, local clusters and the cross-fertilizing effects of research institutions' (Lundvall and Borrás 1997, 39). Thus, several factors contribute to the regional dimension of firms' innovation processes: (i) industrial clusters are in many cases localized, (ii) educational institutions and research organizations are tied to specific regions, (iii) interaction between firms and knowledge providers, knowledge spillovers and spin-offs is often localized, (iv) a common technical and organizational culture may develop to support collective learning and innovation, and (v) regional public organizations have generally been more active in supporting technology transfer and innovation activity in the past years (Tödtling and Kaufmann 1999). Thus, the build-up of different local organizations to create 'institutional thickness' (Amin and Thrift 1994) is emphasized as important in stimulating co-operation, interactive learning and innovative activity.

However, for many firms innovation is a rather internal affair. Reliance on internal competence and lack of trust in external actors are among the main reasons for this (Tödtling and Kaufmann 1999). For other firms co-operation with external partners and inter-firm networking is quite important in order to supplement the internal competence. The networks or systems can be observed at various spatial levels. Thus, firms may innovate successfully without belonging to a regional innovation system as they may find relevant competence milieus, for example, in national or international innovation systems. However, trustful and long-term collaboration in innovation activity with other local firms or knowledge organizations, skilled workers, and a general innovative atmosphere in a region may stimulate the innovation activity of firms, while the absence of, or a weak, regional innovation system may hamper innovation activity in SMEs. In general, possible deficits in the regional innovation system that may hamper the innovation activity of firms can be of three types (this is discussed in more depth in Chapter 3).

First, a regional innovation system may not exist due to a lack of relevant regional actors (i.e. organizational 'thinness'). This points to the fact that not all regions are important units for economic co-ordination. To attain such importance will require sufficient number of firms as well as a knowledge infrastructure in order to enable collective learning. A lack of collective learning may be a deficit particularly in peripheral regions with small industrial milieus and located a long distance from relevant knowledge organizations. However, organizationally 'thin' regions also show that regions differ in their capacity to build up relevant organizations to stimulate firms' innovation activity, depending on their decision-making power, financial resources and policy orientation (Tödtling and Kaufmann 1999).

Second, a regional innovation system may not exist due to a lack of innovation collaboration between players in the region (i.e. a fragmented regional system). Thus, the relevant players may be present, but they do not form a regional system reflecting lack of social capital. Fragmented regional systems show that regions differ in the attitude of local players towards co-operation, which may hamper or advance innovation activity. This follows from the view of interactive learning processes as a basis for innovation activity. Thus, innovation activity nearly always involves some forms of qualitative communication, i.e. interpersonal linkages. The existence of informal institutions facilitates collaboration and the exchange of qualitative information between actors. Thus, 'in networks and other kinds of "organised" market relations, people develop codes of communication, styles of behaviour, trust, methods of co-operation etc. to facilitate and support interactive learning' (Gregersen and Johnson 1997, 482). Such informal institutions mean that firms may enter into different kinds of co-operation without always requiring written contracts, as persons know and follow the same established practices, routines and unwritten rules of business behaviour and rely upon trustful relationships. However, in some regions interaction is hampered, leading to a fragmented system.

Third, a regional innovation system exists, but the system is too closed and the networks too rigid resulting in a 'lock-in' situation. Thus, the other side of cumulative learning and path-dependency that often characterize strong innovation systems is the institutional, social and cultural 'lock-in' of business behaviour. This may be the case if a region historically has had a strong regional innovation system based on R&D institutes and vocational training organizations, with specialized activities dedicated to declining technologies. Such a regional production and innovation system, which has become technologically mature, must upgrade the knowledge base and promote product innovations in order to break path dependency (Cooke 1998). There is also an inherent danger of 'lock-in' in regional innovation systems owing to a homogenization of 'world views' (Grabher 1993a), and these views may become an obstacle to adjustment when technologal trajectories and global economic conditions change. This often creates situations where politicians, labour unions etc. argue for protecting and subsidizing firms in declining industries.

In other regions working innovation systems do exist. However, it is important, analytically as well as politically, also to distinguish between different types of existing regional systems. On the one hand, we find innovation systems that could be called regionalized national innovation systems, i.e. parts of the production structure and the institutional infrastructure located in a region, but functionally integrated in, or equivalent to, national (or international) innovation systems, which is more or less based on a top-down, linear model of innovation (e.g. science parks and technopoles). On the other hand we can either identify networked innovation systems constituted by the parts of the production structure and institutional set-up that is territorially integrated in a particular region, and built up by a bottom-up, interactive innovation model, or innovation networks, which are embedded in the socio-cultural structures of a region, characterized by a 'fusion' of the economy with society (Piore and Sabel 1984), and based on bottom-up, interactive learning (e.g. the traditional industrial district of the Third Italy).

The networked regional innovation system is different from the embedded innovation network due to the systemic dimension of the former, which requires that the relationships between the elements of the system must involve a degree of long-term, stable interdependence. This implies that it is based on system integration and not on social integration. A further consequence of this is that networked regional innovation system cannot be embedded in the community, as embeddedness builds on social integration (Granovetter 1985). However, it is still an example of a bottom-up, interactive innovation model, and, thus, represents an alternative to regionalized national innovation systems. The systemic, networked approach to regional innovation systems brings together regional governance mechanisms, universities, research institutes, technology transfer and training agencies, consultants and other firms acting in concert on innovation matters (Asheim and Cooke 1999). As such it could be said to represent a development towards a 'learning region' understood as a 'development coalition' (Asheim 2001; Ennals and Gustavsen 1999).

The networked regional innovation systems represent a planned, interactive enterprise-support approach to innovation policy relying on close university–industry co-operation. Large and smaller firms establish network relationships with other firms, universities, research institutes, and government agencies. Examples of such networked innovation systems can either be found in regions in Germany, Austria, and the Nordic countries, where this model has been the more typical to implement (Asheim and Cooke 1999), or in later stages in the evolution of industrial districts, which were previously characterized by territorially embedded, innovation networks (e.g. industrial districts in Emilia-Romagna).

Such territory-based regional innovation systems and networks build on different types of knowledge and view of innovative activities compared to the traditional national system of innovation. In addition to the informal, practical and tacit knowledge of 'learning-by-doing' and 'learning-by-using', localized learning processes depend on the important category of disembodied knowledge (in contrast to codified knowledge of a universal character). Different industries, in terms of branch, size and forms of organization, have different requirements with respect to knowledge infrastructures and innovation systems. Locally controlled, traditional SMEs on the one hand may benefit most from networked regional innovation systems or embedded innovation networks, based on an interactive innovation model, while high-tech SMEs and large firms on the other hand may need access to R&D-based knowledge of the linear national innovation systems or transnational (e.g. EU) sectoral innovation systems. Networked regional innovation systems often attempt to link and integrate these different types of knowledge through an interactive university–industry approach.

2.3   CONCLUSIONS

One way of solving the problem of improving the innovative capacity of the small-firm sector of regional clusters, to avoid these firms remaining as firms with a low level of internal resources and competence, is to rely on collective capacity-building by setting up centres for real services and regional innovation systems which could systematically assist firms in regional clusters so that they are able to keep pace with the latest technological development. This could be done either through a networking strategy between firms and public and private agencies, or through public intervention. However, for SMEs to carry out (especially radical) innovations there is often a need to supplement the informal, tacit and localized form of codified knowledge with R&D competence and more systematically accomplished basic research and development, typically taking place within universities and research institutes. In the long run most firms cannot rely only on localized learning, but must also have access to more universal, codified knowledge of, for example, national innovation systems. The strength of the traditional, place-specific and often informal competence and tacit knowledge must be integrated with codified, more generally available and R&D-based knowledge. According to Varaldo and Ferrucci (with reference to industrial districts), 'long-term strategic relationships, R&D investments, engineering skills, new technical languages and new organizational and inter-organizational models are needed for supporting these innovative strategies in firms in industrial districts' (Varaldo and Ferrucci 1996, 32).

Thus, in spite of the important role of close human interaction, place-specific, local resources and regional innovation systems, firms in regional clusters are in need of innovative co-operation and interaction with world-class, national and international competence centres and innovation systems in order to stay competitive. This represents an example of a multi-level approach to innovation systems and knowledge infrastructures. Firms' innovation activity relies on place-specific experience-based, tacit knowledge and competence, artisan skills and R&D-based knowledge. In order for non R&D-intensive firms to be able to acquire formally codified knowledge available from national and international innovation systems, the operation of such systems must be stimulated to become more interactive. In this way, these innovation systems, originally organized according to the linear model, would become more accessible as well as responsive to the individual and collective needs of international competitive, non R&D-intensive firms in regional clusters.

In conclusion, a learning based-strategy of endogenous regional development cannot be applied across the board without some form of public intervention, as well as public–private co-operation to stimulate cluster creation and network formation. However, building up social capital on a regional basis requires some necessary conditions in the socio-economic and socio-cultural structures, and the techno-economic environment, that are more likely to be found only in relatively well-off regions. However, even then a learning-based strategy, for example through the creation or strengthening of regional innovation systems, may be very difficult to accomplish in rural and peripheral areas with little manufacturing industry and traditions, as well as in declining industrial regions dominated by branch plant activities of TNCs, as regionalization trends in such places appear structurally constrained (Asheim and Isaksen 1997; Pike and Tomeney 1999). Many peripheral areas often have too few firms in the same industrial sector or local production system (organizational 'thinness') to constitute a regional cluster, and thus an important condition for local networking and interactive learning is missing. In declining industrial regions it may be difficult in a short-term perspective to bring about the kind of trust and co-operation between a large dominating TNC and local sub-contractors necessary to form regional innovative networks (fragmented regional systems). At least, the first task of a learning-based strategy in these kinds of areas may be to stimulate more collaboration in innovation activities between the large firms and their local subcontractors.

Finally, in the discussion of transfer of experiences from one region to another it is important to distinguish between general and specific factors explaining the formation and development of regions. The more important the specific factors are, the more difficult it is to transfer experiences from one region to another, as specific socio-cultural factors, which are histori-cally rooted in a particular region, cannot be repeated in another region. However, the rapid growth of industrial districts and other specialized areas of production has addressed the perspective of the post-Fordist learning economy on innovation as a socially and territorially embedded, interactive learning process. This constitutes the most significant 'general' lessons to be learned from the particular experiences of various industrial districts. Thus, there are 'general' lessons easier to transfer from one region to another, even if the contingent expression of the experiences can be very specific (Asheim 1994).

PART II

INNOVATION PROCESSES AND
POLICY CONTEXT

3. National and regional contexts for innovation

Arne Isaksen

This chapter gives a general view of the different National Innovation Systems and policy contexts of the 11 regions that host the policy instruments to be evaluated and compared in subsequent chapters. The regions are located in eight countries throughout Western Europe: Austria, Belgium, Britain, Denmark, Italy, the Netherlands, Norway and Spain (see map on page 65). The chapter (i) points to important characteristics, similarities and differences between the national innovation systems and the national 'innovation cultures' in these countries, (ii) describes the various approaches to regional innovation support in the countries and (iii) discusses the typical innovation barriers in the study regions. Thus, the chapter pays attention to the specific institutional and economic contexts of the 11 regions. The chapter forms a basis for some of the analyses in subsequent chapters, as differing national innovation systems and policy contexts influence SMEs' innovation patterns (Chapter 4), the results achieved by different policy tools (Chapter 6), as well as the policy approaches selected in the regions (Chapter 8).

3.1 DIFFERENT NATIONAL INNOVATION SYSTEMS

The first objective of the chapter is to disentangle some of the main similarities and differences between the eight European countries in their approach to innovation policy. Nations differ in important ways as regards their institutional set-ups, socio-economic systems and 'innovation cultures' (e.g. Hodgson 2002). Such variations may lead to differences in innovation policy strategies. What, then, characterizes the innovation policy contexts in these countries?

In accordance with the broad view of innovation (Chapter 2), we also consider innovation policy as a broad policy area. Lundvall and Borrás (1997, 37) define innovation policy as policy 'that explicitly aims at promoting the development, spread and efficient use of new products, services and processes in markets or inside private and public organizations'. According to these authors, innovation policy has wider objectives than those of science policy and technology policy, but incorporates elements of these policy areas. Science policy is concerned with the development of science and the training of scientists, while technology policy involves the use of scientific knowledge in the development of technology, often with the stress on moving into 'higher technology' areas of production. These policy areas focus to a large extent on formal, scientific knowledge, technological innovations and on innovation as a rather linear process.

Innovation policy takes more account of the complexities of the innovation process, focusing on interactions between different types of knowledge as well as interaction between firms and with the institutional infrastructure, including R&D institutes and higher education institutions. Market information and research, and systematic feedback from customers, have also recently been taken into account in innovation policy. Innovation policy builds to a larger degree on the interactive innovation model than science and technology policy. Innovation policy does not primarily focus on the transfer of R&D competence and the stimulation of R&D in industry. A main objective is to foster and speed up learning and innovation processes within firms, and between firms and their environment, where technology transfer may be one of the means (Nauwelaers et al. 1999).

3.1.1   The Significance of National Innovation Systems

Innovation policy arose during the 1980s in Western Europe, mainly at national and EU level, to strengthen the innovation capability and competitiveness of European industry facing increased international competition. Thus, the broad innovation policy is a rather new policy area. Innovation policy lies increasingly at the heart of all industrial policy that aims to raise the competitiveness of national industries. This reflects the view that the strengthening of innovation activity represents a main solution to the demands made on firms, nations and regions by the globalization process through enhancing the learning ability of workers, firms and 'systems'.

Innovation policy is also based on the view that innovation is not necessarily – and not for the majority of small- and medium-sized enterprises (SMEs) – a product from investment in the knowledge-producing sectors only, as seen in the linear innovation model, as discussed in Chapter 2. The policy then stresses moves towards the support of networks and clusters of firms (that may have a regional, national or even a larger geographical extension), and the stimulation of interactive learning among firms and with knowledge organizations. The shift towards the interactive innovation model has, accordingly, increased the importance of the concepts of national and regional innovation systems in policy design.

The interest in national innovation systems reflects a belief that the innovation capabilities of a nation's firms are a key source of their competitiveness, and that these capabilities are largely national and can be built by national policies (Nelson and Rosenberg 1993). Based on the broad understanding of innovation processes, national systems of innovation are seen as systems of interconnected players (like firms, organizations and government agencies) that interact with each other in ways which influence the innovation performance of a national economy, and that this interaction takes place within a specific national context of shared norms, routines and established practices. Thus, at the heart of the concept of innovation systems is the idea that 'the overall innovation performance of an economy depends not only on how specific organizations like firms and research institutes perform, but also on how they interact with each other and with the government sector in knowledge production and distribution' (Gregersen and Johnson 1997, 482).

Innovation systems are open systems, and a specific firm may be part of several innovation systems, be they sectoral, local or national, at the same time. During the last decade, international co-operation in R&D, both technology-based alliances between firms and co-operation with foreign universities and research centres, has increased (e.g. Ernst et al. 2001). The EU's successive Framework Programmes for Research and Technological Development are one important factor in increased international R&D co-operation (EC 1999), and a European innovation system is developing. Although there are many similarities in innovation systems in individual western European countries, some striking differences are seen to exist. Thus, Gregersen and Johnson (1997) regard Europe as a diverse set of national systems of innovation. Firms' innovation performance 'depends on numerous and often country-specific institutional, infrastructural and cultural conditions regarding relationships among the science, education and business sectors, conflict resolutions, accounting practices, corporate governance structure, labour relations etc.' (OECD 1999, 21–2).

The continuous predominance of national innovation systems is demonstrated in an empirical analysis of firms' innovation collaboration in the two bordering regions of Alsace in France and Baden in Germany, divided by the river Rhine (Koschatzky 1999). The study includes 479 manufacturing and business service firms in Baden and 280 in Alsace. While slightly less than 40 per cent of the firms in both regions cooperate with research institutes, none of the firms in Baden has research contacts with an institute in Alsace, and only 7 per cent of the Alsatian firms co-operate with research and transfer institutes in Baden. Spatial distance cannot explain low cross-border co-operation in this case. However, linguistic barriers, differences in mentality and institutional distance matter. Language, laws and diverse national regulations favour innovation co-operation with partners from a firm's own region or nation. Firm managers are often familiar with national R&D institutes due to earlier experience, but unfamiliar with the institutional setting abroad. Thus, in spite of the European efforts for integration and several cross-border initiatives, national innovation systems with their regulations and institutional settings are still important for firms' innovation interactions, and firms in both regions are strongly incorporated in their respective national innovation systems.

What are, then, the main differences in national innovation systems between the eight countries in which the innovation policy instruments assessed in this book are found? Gregersen and Johnson (1997) see national innovation systems in the broad sense as influenced by specific parts of (a) the knowledge infrastructure, (b) the industrial specialization pattern, (c) the institutional set-up, (d) public and private consumer demand and (e) government policy. Variations in these parts will to some extent reflect (f) different national attitudes towards risk, mobility, entrepreneurship etc. This chapter compares the national innovation systems by focusing particularly on differences and similarities between countries as regards knowledge infrastructures and industrial specialisation pattern, the importance of these factors for differing innovation performance, and the ways in which these factors may reflect 'national innovation cultures'.

3.1.2 Differing R&D Systems

The knowledge infrastructure consists of universities, schools, training systems, research institutes etc. These are to some extent oriented towards helping a particular industry or other clients advance their technologies, and they also determine the supply of skills in the labour force. Although the knowledge infrastructure is larger than the R&D system, we start by analysing the amount of resources devoted to R&D in the eight countries focused in this book. As pointed out in Chapter 2, R&D indicators do not cover the full spectrum of innovation processes and particularly the innovation activities of small firms. Research and development is nevertheless of crucial importance in giving firms and countries the capacity to generate, absorb, and diffuse technology.

All the eight countries have a comparatively modest R&D intensity. All are below the OECD average of 2.2 per cent. A low R&D intensity applies to Spain and Italy in particular, signifying a comparatively weak R&D system in these countries.

Governments finance around 40 per cent or more of all R&D in all the countries, except the UK and Belgium, where a very large part of R&D is

Table 3.1   *R&D indicators 1999 (or the last available year), innovation costs 1996*

Country	R&D intensity	Business R&D intensity	Government per cent share of R&D 1995	Innovation costs in manufact.	Innovation costs in services	GDP per head 2001
Austria	1.8	1.1	47.6	3.5	3.0	113
Belgium	1.8	1.7	26.4	2.1	1.2	109
Denmark	2.0	2.0	39.2	4.8	4.7	121
Italy	1.0	0.7	46.2	2.6	n.a.	106
Netherlands	2.0	1.5	42.1	3.8	1.6	116
Norway	1.7	1.3	43.5	2.7	3.5	126
Spain	0.9	0.6	43.6	1.8	n.a.	85
UK	1.9	1.8	33.3	3.2	4.0	103

Notes: R&D intensity is measured as Gross Domestic Expenditures on R&D as a percentage of GDP. Business R&D intensity is measured as business expenditures on R&D as a percentage of business GDP. Government share of R&D concerns government financing of R&D as a percentage of total R&D. Innovation costs are measured as total innovation costs in firms as a percentage of total turnover in firms. GDP per head is measured as GDP per head of population as a percentage of the OECD average. n.a.: Not available.

Source: OECD (2001) and (1999)

financed by business. The large gap between public and private sector R&D investment in Belgium is also reflected in the very small number of R&D personnel working in the government and higher education sector (EC 1999). In fact Belgium is in last place in Europe regarding government R&D infrastructure (Capron et al. 1999). On the other hand the share of expenditure on R&D in the higher education sector financed by government, although declining in the majority of OECD countries, remains very high in Austria, in particular (OECD 1999). To the extent that a high proportion of private funding of R&D is an indicator of technological 'maturity' (Edquist and Lundvall 1993), Belgium and the UK are most advanced in this respect, signifying that Belgian and UK firms have transformed themselves to R&D-based players to a larger extent than usually found in the other countries. Austria, Italy and Spain (with a very low figure on business R&D intensity) are the ones lagging furthest behind as measured by this indicator.

What lies behind the differences in R&D intensity between the eight countries? First, despite a common trend away from the traditional 'mission' in R&D spending on defence and nuclear energy after the end of the cold war, the defence cluster still plays an important role in particular in the UK, explaining part of its high R&D intensity as well as its growth in high technology products (cf. Table 3.3). Defence activities amount to nearly 40 per cent of the total government R&D budget in the UK, and only the United States surpasses the UK among Western countries in R&D spending on defence purposes (OECD 2001). Spain also spends as much as one fifth of its R&D budget on defence purposes, while the other countries use a negligible part of their R&D budgets on defence purposes. Removing this component of the R&D expenditure would set the UK, and even Spain, in a less flattering light than is revealed in Table 3.1.

Second, the figures for R&D intensity partly reflect the industrial and firm size structure of the countries. Notwithstanding the critiques in Chapter 2 about the dubious value of 'high-tech' and 'low-tech' classifications to discuss innovation issues, it should be noticed that a comparatively large share of R&D expenditures is concentrated in a few medium- or high-technology industries (EC 1999), and small firms tend to invest proportionally less in R&D than larger companies. As demonstrated below, Austria, Italy, Norway and Spain are relatively specialized in industries with traditionally low R&D expenditures, contributing to low private R&D spending and hence a relatively high government financing of R&D. In fact, Norway has a level of R&D spending similar to the OECD average when one 'controls' for its traditional industrial structure, i.e. the country is overrepresented with jobs in industrial sectors that generally have low R&D spending. The UK on the other hand, is comparatively specialized in R&D-intensive industries, contributing to a high private R&D spending. There are also substantial differences in the size structure of enterprises between the countries. In Italy and Spain enterprises with less than 50 employees accounted for 69 and 66 per cent of total employment, while the corresponding figure for Austria is 35 per cent and for the UK and the Netherlands 45 per cent (cf. Chapter 1). The significance of large companies is important as long as private R&D spending in several countries is heavily dependent upon a few large companies and the ups and downs in these companies. For instance, in the Netherlands the five largest industrial companies carry out almost half of private sector R&D (Wolters and Hendriks 1997).

What does the R&D indicator tell us about the innovation system in these countries and about the countries' prospects for further economic development? A striking feature in Table 3.1 is the fact that most of the countries have been able for a long time to sustain a high level of income per capita despite a comparatively low score on a traditional technological performance indicator such as R&D intensity. In a country like Norway, the world's second largest exporter of oil, the cause of the high per capita income is to a large extent found in rich natural resources. Italy may be a more instructive case, as (part of) this country represents one of the success stories of post-war economic development, but with comparatively very low R&D spending. This paradox has to be explained by first recognizing that Italy has not one but two very different innovation systems: a small firm network and a core R&D system (Malerba 1993). The small firm networks have developed historically on a local, regional, and vocational basis. The networks consist of a large number of SMEs (in some cases located in industrial districts) operating in traditional industries and specialized in custom-made products and fashion items. The firms interact intensively and share knowledge at the local level in an atmosphere of mutual trust and common understanding. The networks have been innovative, but it is mainly innovation without R&D, and largely insulated from Italian science and technology policy. Innovations are rather the results of experience-based skills, and originate from informal learning by doing, by using, and by interacting (in particular between equipment producers and technology advanced users), and are backed up by specific local institutions and local associations (Garofoli 1992).

The Italian core R&D system is much more recent than the small firm networks. It consists of large firms with R&D departments, small high-tech firms, universities, large public research institutes, and the national government. However, this complex system lacks some qualitative elements needed for effective and successful working (Malerba 1993). This refers to elements such as low R&D capability in several industrial sectors due to relatively few large firms and few small high-tech firms, an uneven level of scientific research in university and research institutes, a lack of technologically progressive public procurement, a lack of co-ordination of public R&D policy, and no tradition of successful industry–university co-operation in research. Thus, the Italian R&D expenditures have not translated into international competitiveness in high-tech industries.

The Italian experience is relevant also for other countries. Traditional SMEs are involved in innovation activities and collaborate in innovation, but these activities are often part of more routine activities and do not register in R&D statistics. Thus, industries may be characterized by considerable innovation, and in a broader sense learning, without a high R&D intensity (Nelson 1993). This is partly reflected in the figures on total innovation costs in Table 3.1 (and in the analysis of 'innovation output' beneath). Figures on total innovation costs give a more correct assessment of innovation intensity than just R&D spending. Innovation costs consist of three main components: (i) R&D costs, (ii) non-R&D innovation costs such as acquisition of patents and licences, expenditures for production design activities, trial production, and market introduction of technological innovations, and (iii) investment in plant, machinery and equipment related to innovation activity.

The UK, the Netherlands and Denmark, having the highest R&D intensity, also have comparatively high innovation costs. This concerns in particular Denmark with its very high innovation expenditures. The UK has high innovation costs in services and the Netherlands in manufacturing. Spain, with the lowest R&D intensity, also has low innovation costs. Austria, in particular, has relatively high innovation costs compared to its modest R&D intensity, i.e. Austrian firms seem to spend a great deal more money on the other innovation activities than R&D. Belgium on the other hand performs relatively better on R&D intensity than on the innovation cost indicator. Thus, Belgian firms in general focus comparatively more on the R&D component in their innovation activity than firms in other countries. This focus may reflect an orientation of policy towards the link between research and industry in this country.

The capability to absorb, diffuse and adapt new technology developed abroad is also crucial, especially for small countries, in addition to the capacity to contribute to the global pool of generic technology by high R&D spending. Thus, indicators like R&D intensity do not catch all (the incremental) innovation of great importance for competitiveness. Nevertheless, the comparatively low R&D intensity and the underdeveloped formal R&D system, as demonstrated in the Italian case, may cause some serious problems in the future in the presence of increasing global competition. Thus, R&D indicators have become a major object of political interest, notably at EU level, where an objective has been set for R&D expenses to reach a level of 3 per cent of GDP by 2010. Small OECD countries are trying hard to improve their relative standing in order to develop more high-tech industries and make R&D a more important part of their innovation systems (Maskell et al. 1998).

3.1.3 Difference in Innovation Results

In assessing the characteristics, strengths and weaknesses of the national innovation systems we also have to consider outputs and results of innovation processes. Outputs have traditionally received less interest than inputs to the innovation process (such as R&D costs) due to a lack of relevant statistical sources. However, Table 3.2 uses some indicators from the Community Innovation Survey (CIS) on the share of innovating enterprises in the eight countries being considered. In this survey an innovating enterprise consists of an enterprise that introduced new or improved products on the market or new or improved processes during a three-year period.[2] CIS puts the same questions to firms in different countries. However, one has to be careful when comparing countries because of differences in data collection methods and sample sizes.

The share of innovating enterprises rises with increasing enterprise size in all countries and for both manufacturing and service industries (cf. Chapter 1). This may partly reflect the fact that large firms often have more types of products or services in their portfolio than smaller ones. Then, it may be more likely that a large firm introduces a significant change in at least one of its products or services in the three-year period than a small firm.

Table 3.2 demonstrates very large differences between the eight countries in their share of innovating enterprises. Denmark, Austria, the Netherlands and the UK have the highest share of innovating enterprises according to CIS both in manufacturing and service industries. Belgium and Spain have the lowest share, and Norway also has a low share of innovating service firms. The differences between the countries are seen in all size classes of firms in Table 3.2. The same result applies to sub-sectors of the manufacturing and service industries (not included in Table 3.2), i.e. the countries with the highest total share of innovating enterprises also have the highest shares in the individual sub-sectors. Thus, the differences in the share of innovating enterprises mainly seem to be an effect of 'nationality', and not an effect of differing firm size and industry structure between the individual countries. Firms in for instance Denmark and Austria are generally more innovative (as measured by the indicators in Table 3.2) than firms in Belgium and Spain, even when considering countrywide differences in firm size and industry structure.

The figures for turnover of new or improved products (the last column in Table 3.2) follow mainly the same pattern as seen in the share of innovating enterprises. Belgium has still the lowest score (14 per cent), i.e. Belgian manufacturing firms have the lowest share of new or improved products from the last three years in their portfolio. Thus, a picture emerges where relatively few Belgian firms innovate. Those which do innovate focus heavily on R&D, possibly to develop radical innovation, which may give few new or improved products, however, having a high potential if the firms succeed in their radical product development. Italy and Spain, on the other hand, have a comparatively high score on the 'new product' indicator. Thus, Italy and Spain, in particular, have relatively few innovating enterprises, but those that innovate succeed well in bringing new or improved products to the marketplace compared to other countries. This may partly reflect the industry structure in these two countries with comparatively many firms producing custom-made and fashion items, where design (as an incremental innovation activity) is important.

A comparison of Tables 3.1 and 3.2 reveals a clear correlation between total innovation costs and the results of the innovation process as measured by share of innovating enterprises. The countries where the firms spend most money on innovation activities also have the highest share of innovating enterprises. The share of innovating enterprises may reflect the effectiveness of the innovation systems in the different countries. Thus, in countries like Denmark, Austria, the Netherlands and the UK, firms generally spend more money on innovation activities and produce more innovation outputs than is the case in particular in Belgium, and to some extent in Spain, Italy and Norway.

Table 3.2   Number of innovating enterprises in different size classes, in percentage (period 94–96 (95–97 for Norway))

Country	Manufacturing sector				Service sector				New products
	Total	20–49	50–249	250+	Total	10–49	50–249	250+	
AU	67	59	73	88	55	54	58	74	31
BE	34	33	34	51	13	11	21	55	14
DK	71	64	76	91	30	24	45	71	21
I	48	44	57	73	n.a.	n.a.	n.a.	n.a.	27
NL	62	54	71	84	36	32	45	71	25
NO	48	39	56	77	22	20	26	50	20
ES	29	21	43	76	n.a.	n.a.	n.a.	n.a.	27
UK	59	54	59	81	40	40	37	55	23

Notes: Innovating enterprises include enterprises which have introduced new or improved products on the market or new or improved processes. Manufacturing sector includes NACE 15–37. Service sector includes NACE 51, 60–62, 64.2, 65–67, 72 and 74.2. New products includes turnover of new or improved products in manufacturing as a percentage of total turnover. n.a.: Not available

Source: Eurostat (2001)

The results in Table 3.2 may also reflect how innovation policy should be designed in these countries. Thus, Belgium and Spain have a much larger share of non-innovating firms than other countries. Then, ideally more efforts could be directed towards raising the number of innovating firms, i.e. helping firms start innovating, than in countries with a much larger share of firms already innovating. In the last group of countries, and in Denmark and Austria in particular, more efforts could be directed towards stimulating more robust innovation projects in the already innovative core of firms.

3.1.4   Industrial Specialization Pattern

The R&D system analysed in section 3.1.2 is one important part of the knowledge generation and diffusion sub-system that is found in any functioning innovation system (e.g. Cooke et al. 2000, 104). The other of the two sub-systems constituting an innovation system is the knowledge application and exploitation sub-system, principally concerned with firms and their vertical supply-chain networks. Differences in national innovation systems and typical innovation performance also appear in this sub-system, i.e. in the production structure and particularly in the specialization pattern of an economy. There are important inter-industry differences in innovation performance, the sources of innovation, the way in which the involved players are connected to each other, and so on. Nations differ in the mix of industries and specialization patterns, and these differences strongly influence the shapes of national innovation systems such as the scientific and technological specialization of their universities and R&D institutions (Nelson and Rosenberg 1993).

Table 3.3 shows the export specialization in some broadly defined manufacturing industries. The table includes the five industries with the largest export specialization indicators in each country. The countries may be divided into three main groups. First, there are four countries – Austria, Italy, Norway and Spain – with their export specialization mainly in the more mature and/or resource-based industries. Thus, these countries have their highest export specialization in: wood products; other manufacturing industries (including furniture); pulp and paper; textiles, leather and footwear; basic metals; non-metallic mineral products; petroleum products; shipbuilding; and food products; all of these are industries belonging to the low or medium-low technology groups. Only few medium–high technology industries are found among those with large export specialization indicators, such as machinery in Italy and motor vehicles in Spain. The four countries all have a comparatively low export specialization in high-technology industries.

The second group consists of the UK only, which has its export specialization industries (except for petroleum products) in some of the most high-tech or R&D-intensive industries. Thus, the UK is in second place, after Ireland, among EU member states in the export of high-tech products (OECD 2001). However, the good performance of British manufacturing export with regard to this indicator is the result of de-industrialization, the heavy decline in areas such as metal-working, and positive restructuring, mainly caused by foreign investment in the electronics industry (Walker 1993). However, the UK's relative strength in pharmaceuticals, chemicals and aircraft is mainly based on indigenous capabilities in British-owned firms.

The third group consists of the Netherlands, Denmark, and to a lesser extent Belgium. Their export specialization industries include their traditional industries such as food products; furniture (or 'other manufacturing'); shipbuilding in Denmark; and petroleum products in the Netherlands. However, these countries also have relatively high exports in R&D-intensive

sectors such as pharmaceuticals (Belgium and Denmark), and in office and computing machinery, and instruments (the Netherlands). At least in Denmark, the 'old' and 'new' specializations are interconnected as some of the strategic competencies in pharmaceuticals can be traced back to experiences and learning in the agro-industrial complex (Edquist and Lundvall 1993). Denmark and especially the Netherlands show very fast increases in high-tech exports during the 1990s (OECD 2001).

Table 3.3   *Export specialization by manufacturing industry and in high-technology industries, 1999*

Country	Industries with a high export specialization and the relevant indicators	High-tech exports
Austria	Wood products (317), Pulp, paper etc. (217), Transport equipment (216), Basic metals (154), Rubber and plastic (145)	14.4
Belgium	Other manufacturing (189), Petroleum products etc. (188), Chemicals (177), Pharmaceuticals (159), Food products (150)	12.9
Denmark	Food products (365), Pharmaceuticals (265), Other manufacturing (182), Shipbuilding (167), Textiles etc. (134)	20.2
Italy	Textiles etc. (286), Other manufacturing (246), Non-metallic mineral products (244), Machinery (183), Transport equipment (167)	10.6
Netherlands	Petroleum products etc. (317), Office machinery etc. (259), Food products (258), Chemicals (133), Instruments (120)	30.3
Norway	Shipbuilding (988), Petroleum products etc. (624), Basic metals (288), Food products (243), Wood products (158)	11.3
Spain	Non-metallic mineral products (244), Motor vehicles (185), Shipbuilding (156), Food products (155), Transport equipment (133)	10.1
UK	Aircraft (183), Pharmaceuticals (155), Office machinery etc. (159), Petroleum products etc. (124), Chemicals (114)	33.5

Note: The export specialization indicator is defined as the share of an industry's export in the country's total export, divided by the industry's OECD-wide share in total manufacturing exports. Values greater than 100 reveal relative export specialization in a country/industry. High-tech export is defined as the share of exports from high-technology industries compared to total manufacturing export.

Source: Based on OECD (2001)

What do the differences in export specialization tell us about differing national innovation systems? The export rate of high tech products may reflect the capacity of a country's innovation system to develop new, technology-advanced products or absorb new and advanced product technology from abroad. An inability in such activities excludes firms and countries from the most rapidly growing sectors and products.

Thus, a weak export specialization in high-technology exports may constitute a long-term structural growth problem in the majority of the countries under consideration. Maskell et al. (1998) argue, however, that it may be advantageous for small countries to increase the competitiveness in their traditional low-tech industries. They also argue that it is fully possible to maintain a high level of prosperity while retaining a low-tech industrial specialization, following a 'low-tech route' to sustained prosperity. This is because many low-tech products and industries are growing and demand technological advancement and continual improvements stimulated by specific inter-organizational factors. The important point in this argument is to consider the complex knowledge base of industries, and not only the R&D intensity of the individual manufacturing sectors. 'The knowledge bases of apparently low and medium technology industries . . . are in fact deep, complex, science-based and above all systemic (in the sense of involving complex and sustained institutional interaction' (Smith, 1999, 1). Thus, as already argued in previous chapters, 'low-tech' sectors may be highly innovative as they are knowledge-intensive from a systemic perspective, and industries may also use scientific knowledge and basic research as part of their knowledge bases. Knowledge links may be indirect, as many flows of technology are embodied in equipment and intermediary products, as well as disembodied spillovers via recruitment of personnel and services provided by the knowledge infrastructure. For example, scientific research and high-technology inputs (in areas such as sonar, medicine, nutrition and robotics) have been of critical importance in developing the Norwegian aquaculture cluster, but firms have received research results mainly in embodied forms, as new or improved equipment, medicine and fodder. Thus, low or medium technology industries are frequently backed up by complex, scientific knowledge bases, or they are part of 'high-tech' national innovation systems.

The Netherlands is the main exception among the countries studied to the 'rule' that small European countries continue their low-tech specialization. Thus, export of high tech products has grown most rapidly in the Netherlands and the UK during the 1990s, and the largest increase is observed in computer and electronics products (EC 1999). These two countries are also the only ones among the eight countries with a surplus in net export of high-tech products, while Austria reveals a very high trade deficit in high-tech products. However, the growth in export of high-tech products does not necessarily reflect national innovation systems more geared towards the development of new generic technologies than before. Thus, the priority given to advanced manufacturing and technological development has often been lower in the UK than in other European countries, reflecting a focus on tradable services based in the City of London, which are predominantly technology users rather than technology producers (Walker 1993). The growing export of high-tech products is partly explained by the Netherlands and the UK being host countries for multinationals producing microelectronic products. Then, growing intra-firm trade (trade between firms located in different countries but belonging to the same multinational group) and re-exportation (foreign multinationals that assemble and re-export final products to other countries) may cause increasing export of high tech products. As a result, growing parts of innovation systems in these countries may be seen as appendages of foreign innovation systems (cf. Walker 1993).

3.1.5   National 'Innovation Cultures'

The national innovation system approach builds on the view that learning and innovations are incorporated in broader societal contexts, particularly in national frameworks of incentives and constraints that are deeply embedded in a set of institutions (Soskice 1999, 102). Three different types of national institutional frameworks that typically support different forms of economic and innovation activity are distinguished by KitSchelt et al. (1999) (building on Soskice 1999). The main distinction is between co-ordinated and liberal (or unco-ordinated) market economies. In co-ordinated economies considerable non-market co-ordination directly and indirectly between companies takes place, with the state playing a framework-setting role. Long-term co-operative relations are stimulated. These economies are further divided into two types covering different European nations. National co-ordination, in which national institutions and governments play a distinct role, has dominated in the Scandinavian countries. In the 'Rhine' capitalist European continent for example, Belgium or Germany, industrial sectors are important in economic co-ordination. Both sub-types of co-ordinated marked economies differ from liberal economies that have little non-market co-ordination between companies, and where the state plays an arm's-length role. The liberal economies are found in Anglo-Saxon countries. Here companies have little capacity to co-ordinate their activity collectively.

The diverse institutional frameworks are seen to have different effects on the strategies company adopt, such as their product-market innovation strategies. Thus, Soskice (1999) argues that co-ordinated market economies have their strength in diversified quality production, which includes relatively complex products and production process, close supplier and customer links, and products that depend on skilled and experienced employees. Liberal economies on the other hand are most competitive in internationally competitive services and industries characterized by radically innovative activities developed in venture capital financed start-ups. In these industries scientific knowledge from national innovation systems based on the linear innovation model is significant. In co-ordinated market economies 'new industries are not easily developed, in contrast to the US and UK economies' (Soskice 1999, 114).

To some extent Soskice's argument is backed by the data revealed above. The UK, as the typical liberal economy among the eight countries, distinguishes itself on some indicators. Thus, the UK has the highest rate of high-tech export, and the country finds its export specialization industries in some of the most high-tech or R&D-intensive industries (Table 3.3). This 'modern' industrial structure reflects partly high defence spending, as industries from which the military procure tend to be R&D-intensive. However, Britain's 'high-tech specialization' may also reflect a relatively high R&D intensity (Table 3.1).

3.2 DIFFERENT REGIONAL CONTEXT AND POLICY APPROACH

This book includes evaluation of some specific policy tools in 11 regions with quite different economic and policy contexts. One aspect of different regional contexts is the differing national policy environment analysed above. Another differentiating factor is the role of the regional administrative level in the design and execution of innovation policy tools in the regions. The autonomy and available funds at the regional level greatly influence where (at which level) and how ('top-down' or 'bottom-up') policy tools are initiated, designed and executed.

However, one has to keep in mind that innovation policy (following the broad understanding of innovation outlined in Chapter 2) is a wide policy area including different administrative levels. Innovation policies consist of three main parts (according to Lundvall and Borrás 1997). The first part is policies affecting the pressure of change, such as competition policy, in order to level the conditions for competition between firms and countries. These policy areas are mainly international, and performed at the EU level. The second part is policies affecting the ability to innovate and absorb change, such as human resource development and the more narrow innovation policy supporting firms' innovation capability. The national level still remains the most important factor in this policy area, controlling the education system for example, though the regional level may play a role. The third part is policies designed to take care of losers in the game of change. Thus, it is supposed that the learning economy, if left to itself, gives rise to polarization between sectors, regions, firms and groups of people. This creates a need for re-distributive policy, a policy area traditionally dominated by the national level. However, the Structural Funds play a growing role in terms of territorial redistribution inside the EU. The third policy area, however, is also included as a prerequisite for innovation. The interaction, co-operation and bottom-up processes that form part of most innovation activity may be difficult to achieve in a socially divided society, and probably more likely to happen in relatively egalitarian societies.

This book evaluates policy instruments included in the second and third policy areas. Actually, these policy areas tend to merge. Inside the 'narrow' innovation policy we have seen a move towards strengthening the regional dimension of policy, associated with the creation of Regional Development Agencies and the formulation of regional innovation strategies. Thus, recent policy discussions focus on the role of the regional institutional environment as an arena to develop a more strategic and customized approach to innovation support (OECD 1998). Inside regional industrial policy we have seen a long-term move from re-distributive or exogenous strategies to endogenous ones (Stöhr 1990). Exogenous strategies focus on the acquisition of enterprises or investments from other areas, while endogenous strategies mean the stimulation of local start-ups and SME growth. Endogenous strategies increasingly focus on the stimulation of innovative activity and capability in local firms and clusters, mainly through obtaining risk capital, upgrading of local skills bases through support for vocational training and incentives for local entrepreneurship, and the transfer of technological know-how to enhance the technological capabilities of SMEs. Partly, these strategies tend to favour bottom-up, region-specific, longer-term, and plural-actor based policy actions (Amin 1999).

In accordance with these development tendencies in innovation and regional policy, a general process of transferring authority from the national administration to the regional level seems to be taking place in most of the countries studied, and very distinctly in Spain and Belgium. The UK has for long been an exception, where the Thatcherite policies led to a reduction in the ability of local government to engage in policy design (Smallbone et al. 1999).

The 11 study regions differ greatly in size from Lombardy's 8.9 million inhabitants to only slightly more than 200000 in the Triangle Region in Denmark. Lombardy in Italy is larger, measured in size of population and

Figure 3.1   SMEPOL study regions

industrial activity, than countries like Norway and Denmark. Besides, the regions differ in their approach to innovation policy, in industrial structure and in their barriers and capacities for change.

Table 3.4 reveals three main institutional and organizational set-ups of government support regarding the role played by the regional level in policy design. The demarcation line between the three regional types is not always clear-cut. However, in three cases the regional level initiates, designs and carries out most of innovation policies. Thus, Belgium is a federal state, where the regional government since 1990 has had the responsibility for technology transfer and for the formulation and execution of innovation policies. The autonomous Spanish region of Valencia also designs and implements innovation policy tools, of which Technology Centres are the most important ones. The regional autonomy and available resources make a strengthening of the regional institutional infrastructure possible, i.e. that more R&D institutes, vocational training organizations, technology centres (as in Valencia) etc. are involved in firms' innovation processes.

In Italy important innovation policy tools are launched by the central government. However, the Italian state is weak, leaving scope for creativity in the manner of regional enterprise and innovation support (Cooke and Morgan 1998), especially to support the small firm network innovation systems. Active regions are key players in the modernization of the regions' industry and the Italian economy more generally. Of our two Italian study regions, Lombardy has been relatively active in introducing 'regional' laws to stimulate innovation processes among SMEs, both through publicly financed service and technology centres and through financial support for innovation projects in firms. Apulia, on the other hand, has relied much more on Community Intervention Schemes, and no one 'regional law' to foster innovation processes has been introduced (Garofoli 1999).

Thus, Apulia belongs to the other extreme in Table 3.4, in which the national (and EU) level designs most innovation policy instruments, disposes the greater part of available funds, and thus carries out the actual work with the tools. While the regions also have some policy instruments, the lack of financial resources, traditions and competence strongly limits the possibilities for autonomous innovation policy at the regional level. In such regions, regional innovation systems will often be regionalized national systems (Chapter 2). Then, intermediary organizations, for example, are national ones or the regionally located organizations are integrated in national or EU-wide networks. The Norwegian TEFT programme (see Chapter 5) offers an example of national intermediary organizations. This programme aims to help regional SMEs to collaborate more frequently with the largest technological R&D institutes in Norway, which are located in some of the country's largest cities. Another example is Tecnopolis in Apulia (also described in Chapter 5), the first scientific park in Italy, founded in 1984. Tecnopolis is a member of the European Network of Business Innovation Centres and since 1993 has been included in the EU Innovation Relay Centres network.

Table 3.4   Organization of innovation policy at the national versus regional level

1) Regionally designed and executed	2) Regional initiatives, mainly based on national (or EU) funding	3) Nationally (or EU) initiated, designed and executed
Wallonia (BE), Valencia (ES), Lombardy (I)	Limburg (NL), Upper Austria	Apulia (I), The Triangle Region (DK), Norwegian regions, UK regions

Denmark, Norway and Britain are seen as examples of mainly nationally or EU-oriented innovation policies. Tools may in various ways be adapted to differing regional circumstances, like the Technology Information Centres in Denmark and the Norwegian REGINN programme that is based on regional innovation analyses. However, the resources and decision-making power are mainly found at the national level. Likewise, the regional policy level is weak in Britain compared with some other European countries. Two of the UK policy tools described in Chapter 5 (the Lee Valley Centre and the Lee Valley Business Innovation Centre) were nevertheless established by initiatives taken by local agencies including local authorities. However, both instruments are heavily dependent upon European and national sources of funding and form part of a wider EU strategy for encouraging and supporting innovation in Europe's more disadvantaged regions.

Regions in the Netherlands and Austria hold an intermediary position regarding the possibilities for innovation policy initiatives at the regional level. In the Dutch study region of Limburg, the most important innovation support for the business sector in the region in terms of volume is the national government's technology policy instruments. However, this policy has been decentralized or regionalized in order to reach out to SMEs, in some respect by setting up 18 Innovation Centres. Within Limburg, European, national and regional policy tools have recently 'merged' into a regional support system, with the Regional Technology Plan (RTP) Limburg at its core. Thus, policy instruments are run by regionally based organizations, with room for adapting the instruments to regionally specific contexts. In Upper Austria, technology centres are the most important support instruments financed at the regional level. However, the lack of financial means in the provinces urges them to co-operate with higher levels (federal state, EU programmes) and to rely on co-financed programmes and private funds (Kaufmann and Tödtling 1999).

A region may be seen as 'a territory less than its state(s) possessing significant supra-local administrative, cultural, or economic power and cohesiveness differentiating it from its state and other regions' (Cooke and Morgan 1998, 64). The 11 study regions constitute regions in this sense to a varying degree. This is to some extent reflected in Table 3.4, where Wallonia, Valencia and Lombardy have the largest administrative and economic power of relevance for innovation policy design. The UK Lee Valley region, the Danish Triangle region and the Norwegian south-eastern region probably have the weakest administrative and political power, as these do not constitute single administrative units. However, Lee Valley and the Triangle Regions are small geographic areas characterized by very distinct economic processes, and in a way constitute 'economic' regions. The Triangle Region, for example, comprises a coherent regional business community and labour market with the stainless steel industry at its core.

3.2.1   The Main Policy Approach in the Study Regions

Another contextual aspect refers to the differing approaches to innovation policies in the study regions or nations. An emerging trend towards more interactive policy approaches is revealed in nearly all the study regions, although with varying intensity. Linear instruments focus on direct R&D aids and transfer of research-based knowledge to firms in order to achieve, in particular, radical technological innovations. Interactive tools, on the other hand, target a wider set of knowledge providers, including other firms, and focus also on other kinds of knowledge (such as experience-based know-how and market information) and other kinds of innovations (such as incremental and organizational innovations). Both types of instruments are seen as relevant, but they often target different kinds of firms (see Chapter 8).

The Italian region of Apulia still relies on rather linear policy tools. The most effective instruments give financial support for the purchase of new machinery, underlining the focus on process innovations and cost-cutting strategies in this region. Cultural and political constraints have also blocked the possibility of introducing schemes targeting local systems and interaction between research institutions and firms in Apulia (Garofoli 1999a).

The emerging Walloon innovation policy can be divided into a 'mainstream' and a 'fringe' part, which are founded on each of the two views on innovation processes (Nauwelaers et al. 1999). The linear approach underlies the 'mainstream' policy with its main objective geared towards transferring scientific and technological advances in industrial processes. The main stimulus for innovation is considered to be the introduction of more formalized R&D; the main channel is through the upgrading of the internal resources in the firm; the 'ideal' innovation is the radical, product innovation; and the technological content of innovative projects should be advanced. The interactive approach has inspired the 'fringe' instruments that constitute an unarticulated and rather fuzzy set of initiatives, and trial-and-error efforts.

The linear approach also dominates the main policy instruments in Austria, the UK and Denmark. Thus, the big national technology and R&D support institutions in Austria still follow the linear innovation model. Similarly, a rather linear approach focusing on the transfer of scientific knowledge to industry is found in the UK Foresight Technology Programme. The support needs in Denmark are also seen in a linear perspective, in which for example the role of the GTS (Approved Technical Services) Institutes is to transform research into innovations in firms.

Some movements towards more interactive support are, however, evident in these countries. Thus, the strategic programme Oberösterreich 2000+ suggests establishing competence centres for specific fields, supporting the formation of industrial clusters, and improving institutions and programmes providing technology transfer services. In Britain some interactive learning conceptualization also influences policy such as the intention to stimulate networking and business clusters.

In Valencia, the Technological Institutes may at the present time be considered as lying in between traditional instruments based on linear innovation models (because they mostly deliver mainly 'off-the-shelf' technologies to firms) and new approaches based on the interactive model (because they are context-sensitive and receiver-oriented). In Norway, a transition from linear tools occupied with the diffusion of scientific knowledge and commercialization of scientific results to more interactive tools, with a stronger focus on demand orientation and context sensitivity is distinct. This is clearly illustrated by the Norwegian tools evaluated in this book, which are all directed towards unravelling firms' real needs for innovation support and fulfilling these needs with a varied set of instruments, which include more than just the supply of research competence (see Chapter 5, and Isaksen and Remøe 2001).

In Lombardy, service centres and technological centres started to work in the 1980s based on a rather interactive understanding of innovation processes. The starting idea was, through public financial support, to help to strengthen the capability of local systems of firms. The centres have also changed their activities due to learning processes that have been taking place for more than a decade. Lastly, Limburg is at the forefront within the Netherlands as far as regional innovation policy is concerned because of the early regional technology plan (RTP) experience in this region. The mainstream approach within the Limburg support system is interactive policy carried out by the Syntens network (the former Innovation Centres).

3.2.2   Typical Regional Innovation Barriers

Chapters 1 and 4 analyse barriers to innovation at firm level. This chapter discusses typical regional innovation barriers, by which we mean hampering factors in the regional industrial milieu, in its institutional set-up, as well as barriers related to the inhabitants' typical attitudes towards innovation and entrepreneurship. The point of departure for identifying regional innovation barriers is the concept of regional innovation systems as discussed in Chapter 2. An innovation system consists of two main groups of players and the interaction between them. The players are first of all firms in the main industrial clusters of the region, including their support industries. Clusters consist of interdependent firms with active channels for business transactions, dialogue and communication (Rosenfeld 1997) both within and outside the region. The other main players in regional innovation systems are knowledge-creating and diffusing organizations, such as universities, colleges, training organizations, R&D institutes, technology transfer agencies, business associations, finance institutions etc. These organizations possess important competence, train labour, provide necessary finance etc. to support regional innovation.

Based on this conceptualization, Chapter 2 identifies three possible deficits in a regional innovation system that may act as barriers in SMEs' innovation activity:

1. Regions may be organizationally 'thin'. In that situation a regional innovation system does not exist due to a lack of relevant players such as local, specialized knowledge organization and/or too few firms in the region. This may, in particular, be the case in peripheral areas.
2. Regions may have fragmented regional systems. Then the relevant players may be present without forming a regional system due to a lack of innovation collaboration. Geographical proximity only creates a potential for interaction, without necessarily leading to dense local interaction.
3. A regional innovation system exists, but the system is too closed and the networks too rigid resulting in a 'lock-in' situation as is often the case in old industrial areas.

In Table 3.5 the 11 study regions are put in one of the three above-mentioned categories. The regions, and especially the larger ones, contain several important industries or clusters, and the regional industries and clusters may suffer from different main barriers. Thus, each region is characterized by several, and sometimes all, of the typical innovation problems, dependent on which industry we focus on. However, Table 3.5 attempts to illustrate each region with one typical barrier.

Table 3.5   Classification of typical regional innovation barriers

Regional innovation barrier	Illustration from the study regions
Organizational 'thinness'	Apulia (I), Northern Norway, Upper Austria, the Triangle Region (DK), Valencia (ES)
Fragmented regional system	Hertfordshire (UK), Lombardy (I), South-eastern Norway, Limburg (NL)
Lock-in	Lee Valley (UK), Valencia (ES), Wallonia (BE)

3.2.3   Organizational 'Thinness'

Organizationally 'thin' regions lack universities and R&D institutes, technology centres, or other important local organisations to stimulate SMEs' innovation activity. The region may also lack industrial clusters, and then firms may find few other local firms to co-operate with.

Five of the study regions illustrate different aspects of organizational 'thinness'. In Apulia, a lack of relevant R&D organizations, in particular, hampers the development of a working regional innovation system. Thus, very few firms in Apulia (13 per cent) find solutions to their technical problems at the local level, while in the other Italian region, Lombardy, firms mostly find help locally (61 per cent). Likewise, research centres and universities are of much less importance in SMEs' innovation activity in Apulia than in Lombardy. A lack of tradition and willingness in inter-firm co-operation and with R&D milieus is also pronounced in Apulia (Garofoli 1999a).

Northern Norway has one university and several colleges and research institutes with competence and research activities of potential relevance for the dominating fish-processing industry, but the region has a lack of knowledge suppliers and players outside the fish industry. Fish-processing firms generally have little co-operation with local organizations when innovating. That may be an important hampering factor, as fish-processing firms in Northern Norway are significantly less innovative than fish-processing firms in the rest of the country. Incremental innovations, and in particular process innovations, dominate in the Northern Norwegian fish-processing industry, and the most important inputs come from equipment suppliers and customers, who are international or in rare cases national (Isaksen et al. 1999).

In Upper Austria the innovation support system has its main deficits in a lack of technology transfer agencies, especially aiming at non-R&D-intensive firms, innovation consultancy, support services targeting commercial aspects of innovation, and risk capital (Kaufmann and Tödtling 1999). In addition the region has no specialized technical university and no contract research organisation. This contributes to the fact that few SMEs actually engage in innovation collaboration with players outside the value chain (or beyond the region or country). Thus, the regional innovation system is highly firm-centred. The most important elements of the technology and innovation system in Upper Austria besides the firms are the six technology centres which have been established recently, and which could take over some of the missing functions of non-profit R&D and basic research in the region. However, hitherto the technology centres have had few connections with SMEs that are not located in the centres, and especially the smallest SMEs, and those SMEs belonging to traditional industries do not use them.

The Danish Triangle Region and Valencia also have few local knowledge organizations. Both these regions are characterized by local production systems of mainly SMEs in mature manufacturing industries such as footwear, textiles, ceramics, toys and furniture (Valencia), and the food-processing equipment industry (the Triangle Region). The R&D component is rather weak in the innovation system in both regions. Thus, the Triangle Region has a lack of higher education institutions and research institutes (Christensen et al. 1999). However, these may to some extent be found in nearby areas. The Valencian Technological Institutes have paid little attention to R&D (except for the Ceramic Institute). The institutes provide services 'off the shelf', such as standard testing etc. However, more R&D competence is necessary in the regional network of firms due to increased competition resulting from economic integration in the EU and globalization processes (Vázquez Barquero et al. 1999).

3.2.4 Fragmented Regional Innovation Systems

The second typical innovation barrier points to the fact that a working innovation system does not exist automatically, even if all the relevant players are present in a region. The players also have to interact, e.g. the firms have to make use of the regional knowledge organizations. Thus, the second barrier indicates that regions differ in the attitude of local players towards co-operation, which may hamper or advance innovation activity.

Four of the study regions may illustrate aspects of fragmented regional systems. The outer metropolitan area of Hertfordshire near London is typically viewed as a dynamic and growing economy. The region is a favoured location for new investments in R&D-intensive industries, in particular aerospace and pharmaceuticals, and several major firms have set up both R&D facilities and production plants there. The area has the third largest number of R&D workers in Britain, and it is located near a dense location of leading universities, research institute laboratories and hospitals in the south-east of England. Thus, the elements forming a strong regional innovation system are present.

However, the area may suffer from some general problems related to UK industrial development. A general characteristic of British industrial culture relates to the lack of a tradition of inter-firm networking and low trust, which is in accordance with Soskice's (1999) view of the UK as a liberal economy typified by little non-market co-ordination between companies. Little inter-firm networking may hamper the development of working innovation systems, both at the regional and national level, as innovation interaction between firms is a prerequisite for developing an innovation system. Little inter-firm networking may be 'symptomatic of an industrial economy which has prioritized low cost over quality and where firms place greater emphasis on short-term relationships than on negotiation and longer-term co-operative relations with other firms' (Smallbone et al. 1999, 21). In addition, UK regions to a large extent lack the organizational infrastructure to support the development of regional innovation capacity, a further prerequisite in order to develop regional innovation systems.

The Italian region of Lombardy is, like Hertfordshire, a dynamic part of the national economy. Lombardy has a considerably varied industrial base, and is the region in Italy that spends most on R&D, focusing on applied research that contributes to an efficient and competitive R&D system. However, Lombardy seems to suffer from the same innovation barrier as Italy as a whole, i.e. a lack of interaction between the small firm network systems and the core R&D system (Malerba 1993). Thus, innovative firms in Lombardy sometimes have no or few links to the regional knowledge infrastructure, and it is generally difficult to involve universities and research centres in the region in innovation activity in local firms. The lack of interaction is very often found in cultural barriers (Garofoli 1999)

Studies of regional clusters in south-eastern Norway also point to the existence of fragmented regional innovation systems. However, regional resources form an important basis for innovation activity in the regional clusters, i.e. firms rely on regional resources when innovating. These resources include unique combinations of knowledge and skills among the labour force and specialized suppliers. However, many firms in the clusters have 'grown out' of their home region when it comes to more radical technological development. This reflects partly a lack of relevant competence in the regional R&D system. Some firms are world leaders in their niches, and they have to co-operate with the 'best' R&D milieus, which they find at the national and international level. Also many SMEs have important external contacts. These considerations point to the relevance of a multi-level approach, rather than just focusing on the regional or national level, when analysing innovation processes, as firms exploit both place-specific resources and external, world-class knowledge respectively when innovating (Asheim and Isaksen 2000).

The Dutch region of Limburg also suffers from a relatively poor public research and development infrastructure as far as technical science is concerned. Actually, the R&D centres of two large multinational firms are the prime players in Limburg in creating 'new' research-based knowledge (Nauwelaers et al. 1999). However, the RTP study from the province of Limburg points to a lack of innovation interaction as a more relevant innovation barrier in the area. Considering the fact that the major players are company R&D centres, which have to deal with issues like secrecy and proprietary knowledge, increasing innovation collaboration may be difficult to achieve. The RTP study states that 'the availability of technology as such can hardly be considered a problem. The problem lies more in the access to knowledge. Companies, especially SMEs, are not sufficiently aware of the importance of the knowledge infrastructure. Moreover, the knowledge centres and the intermediary organizations have not yet succeeded in taking know-how to the company effectively' (RTP-Limburg 1996, 83).

3.2.5 'Lock-in' of Innovation Systems

The third typical regional innovation barrier points to the fact that too strong ties may lead to 'lock-in' situations. Several of the study regions have elements of 'lock-in' due to a history of dynamic industrial development. We illustrate 'lock-in' by means of the two old manufacturing regions of Lee Valley and Wallonia. Historically, London's Lee Valley has provided a social, economic and institutional environment to encourage and facilitate innovation and entrepreneurship. Thus, at one time Lee Valley was London's foremost industrial area with an impressive number of successful manufacturing companies. However, it is highly questionable whether the area is continuing to provide an enabling environment for the 'new' types of economic activity.

Over the last two decades, Lee Valley has experienced considerable industrial decline and associated high levels of social deprivation, and the area has been an EU assisted area, receiving Objective Two status for the 1994–99 period. A certain 'lock-in' in Lee Valley is indicated in an evaluation of the Small Firms Merit Award for Science and Technology (SMART) scheme (Smallbone et al. 1999). SMART targets R&D-based firms and entrepreneurs based on an annual competition (Chapter 5). The take-up of SMART is much lower in Lee Valley than in the other UK study region of Hertfordshire, signifying that the kinds of firms and entrepreneurs that develop new science- and technology-based products are relatively seldom found in Lee Valley. Several universities have become partners in regeneration policy initiatives. However, the SMART evaluation indicates that firms and entrepreneurs in Lee Valley often lack the competence in R&D-based innovations, and the kinds of firms and entrepreneurs that are most able to compete for SMART funds are not 'raised' in the industrial milieu found in Lee Valley.

Wallonia has experienced the same kind of industrial development as Lee Valley. Thus, Wallonia was one of the first regions of the European continent to industrialize, based on industries such as coal and steel, textile and glass. Wallonia is still heavily dependent on mature industries in declining sectors, but the region has been involved in a deep restructuring process during the last few decades. Still, much remains to be done, as evidenced by the Objective 1 status that was given to part of the region until the end of the nineties. The region has developed a considerable research and higher education infrastructure, and highly innovative firms have clustered around universities, and some of Wallonia's science parks are internationally renowned. However, the former dependence of large companies may still be reflected in a general 'lock-in' of entrepreneurial spirit in the region, and a 'sub-contractor culture' in SMEs (Nauwelaers et al. 1999). The region has also been less successful in networking within industry, and between industry and the knowledge infrastructure. The Walloon regional innovation system also suffers from a relatively poor business service sector, which may uphold a 'lock-in' situation.

3.3   CONCLUSIONS

This book includes comparative evaluations of almost 40 different innovation policy tools aimed at SMEs drawn from 11 European regions. This chapter has considered the specific institutional and economic conditions in the regions, as these form a basis for some of the analyses and policy discussions in subsequent chapters.

One contextual aspect refers to the differing national policy environments and innovation systems. All the eight countries in question have a comparatively modest R&D intensity. Denmark, the Netherlands and the

UK are closest to the OECD average. A low R&D intensity applies to Spain and Italy in particular, signifying a comparatively weak R&D system in these countries. Regarding industrial specialization patterns the eight countries may be divided into three main groups: (i) Austria, Italy, Norway and Spain have their export specialization mainly in the more mature and/or resource-based industries, belonging to the low or medium-low technology groups. (ii) The export specialization industries in the Netherlands, Denmark, and to a lesser extent Belgium include both their traditional industries and relatively high exports in some R&D-intensive sectors. (iii) The UK alone has its export specialization industries almost entirely in some of the most high-tech or R&D-intensive industries, reflecting a national institutional framework geared more towards the development of radical innovations and the early growth of new industries than the other countries in question.

The chapter shows that the regions hold quite different innovation systems on various aspects. The regions diverge in their authority to design and carry out policy instruments, in their approach to innovation policy and in their innovation system barriers. In three of the regional cases – Wallonia, Valencia and Lombardy – the regional administrative level initiates, designs and carries out most of the innovation policies. The other Italian study region, Apulia, as well as the Danish, Norwegian and British regions are seen as examples of mainly nationally- or EU-oriented innovation policy, in which the national or EU level designs most innovation policy instruments and disposes of most of the available funds. Regions in the Netherlands and Austria hold an intermediate position regarding the possibilities for innovation policy initiatives at the regional level.

Another contextual aspect refers to the differing approaches to innovation policy in the study regions or nations. An emerging trend towards more interactive policy approaches is revealed in nearly all the study regions, although the linear approach is still the dominant form of intervention. Limburg seems to have the most interactive support system, while Apulia still relies on linear instruments.

The study regions also illustrate different typical regional innovation barriers concerning SMEs. Some regions are organizationally 'thin' with a lack of relevant knowledge providers and/or few firms to stimulate SMEs' innovation activity. Apulia, Northern Norway, Upper Austria, Valencia and the Danish Triangle Region illustrate different aspects of organizational 'thinness'.

Four other study regions may illustrate aspects of fragmented regional innovation systems. Fragmented systems point to the fact that linkages between players are not well developed even if all the relevant players are present in a region. The UK region of Hertfordshire, Lombardy, south-eastern Norway and Limburg display aspects of fragmented regional systems. The third typical regional innovation barrier points to the fact that too strong ties may lead to 'lock-in' situations, a situation found particularly in the two old manufacturing regions of Lee Valley and Wallonia.

4. Innovation patterns of SMEs

Alexander Kaufmann and Franz Tödtling

In Chapter 4 of this volume the regional case studies of the SMEPOL project are analysed with the focus on the innovation activities of SMEs and the problems they are confronted with. From these innovation patterns implications for policies aiming at supporting innovation in the SME-sector are deduced. These results are intended to provide a fundamental guideline on which aspects SME innovation support instruments should focus in order to be effective. This and the evaluation of the investigated instruments in Part III will deliver the empirical base for drawing policy conclusions in the final chapter of this book.

This section concentrates on those basic findings concerning SMEs' innovation activities that are the most important starting-points for support instruments: types of innovations and the level of innovativeness, the role of innovation in the set of strategies of SMEs, the nature of the innovation process, the resources dedicated to innovation, the external relations in this process, and the problems constraining innovation. Using the results of the regional case studies, a distinction will be drawn between findings that are general in nature, and findings that are more region-specific.

4.1 THE INNOVATIVE PERFORMANCE OF SMEs

In the SMEPOL project the term 'innovation' is used in a very broad sense. It covers any modification or improvement of products or services, any kind of change of a firm's range of products or services, and upgrading as well as adopting or developing new technologies. We do not deal with organizational innovations in this chapter, however. Different levels of innovativeness are determined mainly according to the perspective of novelty. If a product innovation is new only for the firm but not to the market, then this is called 'less innovative' than products or services which are new to the market. Products new to the market (major innovations) are functionally different from existing products on the respective market and they face less competition as a consequence. Any technical improvement, redesign or new product added to the product range which has to compete with similar products of other firms is therefore considered a minor innovation. (The frequency of product innovations in certain regions investigated in the SMEPOL-project are presented in Table 4.1.)

Another frequently used differentiation is 'incremental' versus 'radical'. These terms are difficult to operationalize, because it is hardly possible to draw a line between changes in existing products without replacing them and the introduction of products based on a new concept or performing new functions. Therefore we will not use these categories in order to classify innovation. It has to be considered that nearly all innovative activity takes place within the range defined above. Innovations which are so fundamental that they open up new technological trajectories enabling the establishment of a whole family of applications leading to several linked markets for products and services (e.g. the personal computer) are so rare that this category (the truly 'radical' innovation) is of little use as orientation for innovation support policy.

Most SMEPOL studies show that SMEs are less innovative than large firms (Upper Austria, Norway, Wallonia, Limburg), but there is considerable variation between types of SMEs. According to the results of the Upper Austrian survey this difference applies to any kind of product innovation (including modifications of already existing products) as well as the more advanced innovations (which are new to the market).

Overall, the SMEPOL case studies are in line with evidence that SMEs are generally less innovative than large firms (Eurostat 2001; Craggs and Jones 1998; Tödtling and Kaufmann 1999; Cooke et al. 2000). Nevertheless, there are also findings (Pavitt et al. 1987; Acs and Audretsch 1990) leading to the opposite conclusion (see also Chapter 1). Partly this ambiguity is due to the heterogeneity of the SME sector regarding innovation. Innovativeness ranges from firms without any innovative activity to highly dynamic high-tech companies able to introduce radical innovations. Innovativeness varies across size-classes, industries and regions. Examples are the plastics sector in Upper Austria which is the most innovative group of firms in the survey whereas the plastics industry in Austria as a whole belongs to the less innovative industries. In Wallonia, too, the comparatively more innovative industries are not only the typical high-tech sectors like electronics, but chemicals and rubber/plastics. In the chemicals industry of Wallonia biotechnology and pharmaceuticals are responsible for the good innovation performance.

In general, the service sector seems to be less innovative than the manufacturing sector. This is confirmed by the Upper Austrian, the Walloon, and the Limburg results. Contrary results can be found in the case of London. Excluding simple modifications, within London the business service firms (computer, telecommunications and other business services)

Table 4.1 Frequency of product innovation in percentage of the firms

Region	Type of firm	Any kind of product innovation	Advanced product innovation
Upper Austria:		(including modification)	(product new to the market)
	All SMEs	82.9	45.0
	1–9 employees	70.9	45.5
	10–49 employees	87.5	47.5
	50–249 employees	95.0	45.0
	Large firms	96.9	60.9
London:		(including modification)	(innovative compared to other firms)
	All SMEs	84.0	58.0
	1–9 employees	80.0	46.0
	10–49 employees	93.0	66.0
	50–249 employees	90.0	60.0
Denmark:		(product new for the firm)	(product new to the market)
	All SMEs	38.2	27.9
	1–9 employees	26.5	22.1
	10–49 employees	46.9	32.7
	50–249 employees	57.9	36.8
Valencia:		(including incremental improvement)	(completely new product)
	All SMEs	84.0	27.0
Lombardy:		(any innovative product)	(product new to the market)
	All SMEs	76.4	41.6
Apulia:		(any innovative product)	(product new to the market)
	All SMEs	40.7	16.7

Source: SMEPOL surveys

are more innovative than the investigated manufacturing sectors (food processing, computer and machinery equipment manufacturing, electrical and instrument engineering). In the outer metropolitan area they are almost as innovative as the engineering industries and more innovative than food processing.

As to be expected, firms applying higher levels of technology are, in general, also more active in product innovation. Nevertheless, there are exceptions where low technology and frequent innovation activity are no contradiction. Examples are industries like textiles and footwear in Valencia which are highly dependent on rapidly changing customer needs and fashions and which are therefore frequently innovative, even if they are not technologically advanced. Most SMEs of these industries can be considered as innovative if we include modifications, and 70 per cent to 86 per cent have introduced new products. These innovations, however, are predominantly incremental, often (especially toy firms) imitative. Imitation seems to be a very important form of innovation also in the Danish stainless steel industry.

In Wallonia and Norway there is a strong positive correlation between the frequency of general product innovations and size of the SMEs. In Upper Austria, too, this correlation is positive, but weaker. In Denmark, it is especially the smallest firms with less than 10 employees that are lagging behind. Focusing on more advanced product innovations the positive correlation between size and innovativeness is more ambiguous. In Upper Austria there is almost no difference in the frequency of products which are new to the market between SMEs belonging to different size classes. In other regions, however, advanced innovations are more frequent in the case of larger SMEs. In Wallonia larger firms are more active than smaller firms with regard to the introduction of products which are new to the market. In Denmark, too, there is a positive correlation between size of SMEs and products, which are new to the world or national market or the industry. In London, innovations which are new to the market, are least frequent in very small firms with less than 10 employees.

Regarding policy, these findings suggest that support instruments must be designed in targeted, flexible, and at the same time comprehensive ways in order to meet the needs of very different SME clients. Support needs to be targeted regarding the specific needs and the ability of firms to use the support successfully. It implies that the support services offered have to be designed in a way adequate to the needs of different types of SMEs. In addition, it must be avoided that support tries to reach as many SMEs as possible disregarding the expected return in terms of increased innovativeness and competitiveness. Standards and acceptable limits, which supported firms have to achieve concerning innovative and competitive performance, are required. On the other hand, no SME should be excluded because of structural or sectoral features. The instruments should be open enough to reach 'high-tech' as well as 'low-tech' firms, to cover the whole range from the smallest to medium-sized firms, and to be useful both for manufacturing and service firms. Flexibility refers to the necessary adjustment of the instruments, which has to be done continuously based on a routine process of evaluation. Comprehensiveness refers to the scope of support functions offered to SMEs which should cover more than technical issues and should target also innovation-related activities before and after the development and introduction of new products like market and technology information and the commercialization of new products.

4.1.1   Innovation Strategies of SMEs Tend to have a Defensive Character

Although innovations strategies and objectives are rarely formally stated, in fact SMEs often specialize on market niches and focus primarily on quality advantages. According to the SMEPOL case studies small changes of products as well as the innovative redesign of products are, besides cost-cutting, the usual innovation goals.

In Upper Austria the most frequent innovation strategies are specialization on niches and quality advantage. This behaviour is due to the limited scope of products and markets of SMEs. For large firms it is easier to engage in offensive strategies like entering or opening up of new markets. Nevertheless, large firms too pursue primarily defensive innovation strategies with cost-cutting as the predominant objective. Cost-cutting is also a typical innovation objective of small firms, however. In the Danish stainless steel-cluster cost-cutting-related reasons are predominant for SMEs to innovate. Nevertheless, new technical possibilities and the wish to extend the product range also rank high in importance. In Wallonia there is little difference in the frequency of offensive strategies between SMEs and large firms; defensive strategies are even less frequent in the case of SMEs. Similar to Upper Austria, for large firms cost aspects are more important than for SMEs.

The defensive character of SME innovation is not an Austrian specificity. Also, according to the London survey, the most frequent basis of product innovation, except for services, is rather defensive: 'innovative redesign of a traditional product'. In the other regions the results are more ambiguous. In the Norwegian and the Walloon cases it was not distinguished between the goal of opening of new markets and the goal of increasing market share. Only the first objective is undoubtedly offensive. It is not clear, therefore, how many firms concentrate on raising their share in their old market, relying on quality improvements and price competition.

In Norway the latter interpretation seems to be more appropriate, because more defensive strategies like improving product quality and reducing labour costs rank first and third within the group of innovative SMEs. The extension of the product range is only the fourth frequent objective of innovation. Opening of new markets and increasing market share become more frequent with firm size. In general, offensive strategies are more frequent in R&D-intensive SMEs, defensive in firms not engaged in R&D. In the case of Wallonia the offensive character of innovation might be more important. The diversification of the product range ranks second as innovation objective behind opening of new markets and increasing market share.

The London survey shows that innovation is part of almost all SMEs' strategies, but more frequently in an informal rather than formal way, as already argued in Chapter 1. In Lombardy, innovation is not a top competitive factor, at least formulated as an explicit strategy. In Apulia innovation is hardly seen as a competitive factor at all. Nevertheless innovation is implicitly part of the most frequent competitive advantages like product quality and problem-solving capability. This is further confirmed by the fact that by far most of the firms in both regions intend to introduce new products. In contrast to Apulia, for more than half of the firms in Lombardy new products are a priority task.

The implications for policy are that instruments should, first, raise the capabilities of SMEs to search for innovation-relevant information and, second, offer market and technology information showing windows of opportunity to SMEs. To improve the monitoring skills in SMEs is one part of the general support strategy to raise the capacity to innovate and to exploit the realized innovations (see also the following sections). But even if firms are well trained to access and use information from a wide range of sources, it is still helpful if firms can use edited and organized information. Providers of such information services have to do this in a way that very small and low-tech firms are also able to use this information. First, it is necessary to distribute this information proactively. Most SMEs will not use information presented without active promotion. The barrier of personal energy required to collect information should be kept to a minimum. Second, the information about ongoing market and technology developments and their potentials has to be usable for representatives of SMEs. Concrete examples and realistic scenarios should demonstrate that certain innovation projects are realizable, and how this has to be done by other firms. For this purpose it is necessary to target information on certain types of SMEs, distinguished by technological level, industry, position in the production process, size, and similar characteristics, so that it is as close to their perception of economic and technical reality as possible. If the information presented is too abstract or the realization and application seems to be too remote, SMEs will not be stimulated to engage in innovation activities on their own.

4.1.2   Process Innovation is often Independent of Product Innovation Aiming at other Objectives such as Cost-cutting or Flexibility of Production

The introduction of process innovations and the adoption of new technologies can be independent of or related to product innovations. Particularly in the case of incremental changes or modifications of products, process innovations are often not required. If the product innovations are new for the firm or appearing the first time on the market, then, usually, new production technologies, adopted from outside or developed by the firm itself, will be necessary. Nevertheless, process innovation can also be completely independent of any product innovation activity. Usual reasons for new technologies without changes in the product range are cost-cutting, improving delivery, flexibilization, broadening/speeding-up of provision and collection of information etc.

The introduction of new technology to enable the development of new products occurs only in specific segments of the SME-sector. In Upper Austria, process innovations are generally less frequent than product innovations. However, the difference between the frequency of product and process innovation decreases with firm size. Very frequent reasons for process innovations are not linked to new products like improving of quality and productivity. Nevertheless, 30 per cent of the innovative SMEs required new technologies for their product innovations. Most of those SMEs, which have introduced products new to the market, needed new technologies too. Most technologies are adopted; firms developing technologies on their own are less frequent.

Less correlation between product innovation and process innovation was found in the case of London. Most process innovations aim at the expansion of production and the improvement of quality. Less frequent are new technologies which are an integral part of the production of new products or services. This latter reason was found most frequently in the business service sector, least in computer and machinery equipment manufacturing as well as in electrical and instrument engineering which focus more on quality improvement. A gap between new technologies and product innovations is further confirmed in the special case of the Internet. The use of websites is quite widespread already, more frequent in larger SMEs and the service sector. But only a minority of these firms uses the website for marketing purposes, which shows that the potential of this new ICT is not sufficiently utilized as a means of strengthening product innovativeness.

Also, in Italy only few firms (about 12 per cent) invested enough in new machinery to be able to introduce new products. The most frequent reason for investing in machinery is the extension of production both in Lombardy and Apulia. In Lombardy 'new technology for new products' is just as frequent a reason as 'improvement of quality' or 'substitution of old machinery'. In Apulia other reasons are comparatively more important: increasing labour productivity, lowering production costs, and quality improvement.

No relation between process and product innovation could be found in the case of the Norwegian fish-processing industry. In this industry process innovation is clearly more important than product innovation. There is little activity concerning product innovation. Process innovation in mature and low-tech sectors like fish processing means usually the adoption of existing technology, often from outside the industry. The main reason for new technologies is to improve efficiency and productivity and to increase flexibility in order to be able to deal with a rapidly changing supply of fish and demand of consumers.

Other cases where the introduction of new technologies is the dominant form of innovation are the investigated local production systems of Valencia. There imitation of leading firms is the main process innovation activity. This applies in particular to small firms, larger firms are more often innovating on their own. A large part of process innovations concern the introduction of computers for data processing, automation, testing, and control. The specific value chain in each production system influences strongly the type of and motivation for innovation. In the ceramics sector certain stages of the production process have been outsourced to newly established local firms specialized in the major process innovations. In textiles innovation is predominantly aiming at improving quality, increasing speed and flexibility. New technologies, however, are mainly adopted from specialized machine builders located abroad. In the manufacturing of footwear the major process innovations come from other industries (chemicals, for example). This is similar in the toys sector. In general, in these sectors process innovation seems to be slightly more important than product innovation.

Policy should avoid supporting new technologies in general, because there are several reasons for process innovation. Often process innovation has a rather defensive character such as cost-cutting. If the primary objective is to support offensive, dynamic firms, then instruments should concentrate primarily on those new technologies which are necessary for product innovation, concerning quality improvements as well as completely new products. A further aspect of support is to enable firms to fully exploit the potential of new technologies for product innovations.

4.2   CHARACTER OF THE INNOVATION PROCESS

4.2.1   SMEs Innovate with a High Resource Intensity, in Particular Regarding Human Resources

Although only few resources are available for innovation in most SMEs in absolute terms, SMEs typically dedicate larger shares of their financial and human resources to innovation than big companies. This is especially true of human resources. In Upper Austria SMEs dedicate 11 per cent of their turnover and 15.6 per cent of their manpower to innovation on average. For large firms the mean values are 10.3 per cent and 8.3 per cent respectively. This is also confirmed by results for Limburg and the Danish Triangle region where the financial innovation intensity clearly increases for the smaller firms. As far as R&D is concerned, the same tendency of higher resource intensity in smaller firms can be observed in Norway. The higher R&D intensities (4 per cent of sales and more) are more frequent in the case of SMEs with less than 50 employees. It has to be considered, however, that the 'R&D' activity as such (disregarding intensity) is more frequently performed in larger firms. SMEs, on the other hand, are more likely not to do any R&D at all. SMEs, in relative terms, rely much more on manpower compared with financial means than large firms. Normally, in small firms a higher proportion of persons are involved in innovation than in large firms where it is possible to organize some innovation-related functions in full-time jobs or separate departments.

According to the survey results from Upper Austria, Denmark, and Lombardy, few SMEs have a share of sales based on innovative products of more than 50 per cent. The share tends to be higher in the case of smaller firms. Many of them are young and rely on one or a few innovative products or services. Nevertheless, it is usually necessary to have a significant range of products which are easy to sell ('cash cows') in order to be able to spend sufficient funds on innovation activities. This is certainly a particular problem for the smallest SMEs.

In designing support instruments, policy will have to focus especially on manpower available for innovation in order to be effective. This applies, in particular, to capabilities of innovation management (organization, time, strategy), market research, technology monitoring, and financial management (especially risk capital funding). This is to a large extent 'pre-R&D-project support' aiming at building up capacity to innovate. To improve human capital it is not only the financial aspect that support should target, for example by (co-)funding R&D personnel. Furthermore it will be necessary to increase the availability of adequately qualified personnel. The concept of 'innovation assistant' might be a model. In addition, the training and education system might have to be upgraded or extended in order to be able to 'produce' a sufficient number of people with the required skills.

4.2.2 Research is a Less Frequent Activity in the Case of SMEs than Large Firms

Partly due to the limited resource base (see Chapter 1), SMEs are less often engaged in systematic research and development activities than large firms (see also Cooke et al. 2000 and the latest Community Innovation Survey, Eurostat). This finding is confirmed by the Upper Austrian and the Danish SMEPOL results. Of course, there are highly research-intensive SMEs, but in general, SMEs are confronted with serious size-specific barriers restricting the potential to do research. One barrier concerns manpower. Often key persons are preoccupied with day-to-day work and lack therefore the time to engage in innovation projects. Another problem concerns the limited financial capacity of SMEs. It is far more difficult for them to engage in research and development of products which are many years away from commercialization than for large firms. The product range of SMEs is small and they usually cannot substitute for the lack of sales and profits through other products ('cash cows') to the same extent as large firms. In addition, product- and innovation cycles are short, partly due to the incremental character of innovation. Within the Danish stainless steel sector 78 per cent of the firms have a time period for product development of not longer than two years. According to the London survey, lower technology SMEs from the food processing sector and services have most often short innovation cycles between idea and launch. For most of them the time period is less than one year, reflecting the incremental character of their innovation activities. The higher-technology industries, computer and machinery equipment manufacturing as well as electrical and instrument engineering, have clearly longer cycles (for 50 per cent between two and five years, and for 15 per cent over six or more years).

In order to increase the research activities in the SME sector innovation policy will have to apply a diversified set of instruments. At first it is necessary to mediate and stimulate interactions with universities and research organizations. Due to the fact that many SMEs do not have adequately qualified personnel to communicate with science, it is furthermore important to support the firms' ability to employ such persons through financial support or mobility schemes, for example. Finally, there is a need for long-term funding of innovation projects including market introduction and penetration and funding for long-term R&D projects. Possible ways are to provide venture capital or to mediate and support the access of SMEs to private venture capital funds or providers of risk capital. But it is not only the lack of available or accessible risk capital that poses problems to finance innovation project; often SMEs are not ready to make use of this type of funding. Many entrepreneurs fear to lose independence or control of the firm making it difficult for them to engage directly in venture capital constructions. It might be a reasonable support approach, therefore, to offer the possibility of a stepwise change from direct support for innovation projects to risk capital funding. An interesting model might be the so-called 'mezzanine capital' which is a loan similar to equity capital. There is no or a very low interest rate, no participation in capital expansion, but, after a certain period of time, the loan has to be substituted by a private investor. This gives the firm some time to get accustomed to private risk financing.

4.2.3 In SMEs there are only a Few Persons Able to Act as Potential Nodes for Linking with External Innovation Networks

The low capacity of many SMEs in terms of manpower to establish and maintain interfaces to innovation networks applies both to the search for and collection of innovation-related information and the co-operation in innovation projects. In Denmark the lack of experienced employees dedicated to product development is one of the main barriers for collaboration, the other is the lack of time. Market research is very rare in Upper Austrian SMEs (17 per cent), independent of size or technological level. In London the introduction of new sources of market information is also rare, but there is a clearer correlation with size and technological level; larger firms and higher technology SMEs are more engaged. Therefore the danger of lock-ins is greater in the case of SMEs than large firms. According to Norwegian data, for example, there is a clear correlation between the size of a firm and the number of different types of innovation partners. Larger firms can establish relations with a broader range of co-operation partners than smaller firms. In some of the studied groups of SMEs in Norway, particularly the regionally isolated SMEs, the problem of too little time for innovation due to routine and administrative work became very clear.

Innovation policy should try to strengthen the boundary-crossing functions, as a way of improving the capacity of the relevant people to interact with the firm's environment as well as by increasing the number of intra-firm nodes, for example by (co-)funding of innovation assistants and consultants. But it is also possible to offer indirect support in order to increase the available time for innovation of key persons. For this purpose it is necessary to ease the access to innovation-related information (see also the previous section). It must be clearly structured, regularly updated, and presented in an interesting and reasonable way, so that SMEs can easily and rapidly make use of it. Information technologies, the Internet in particular, make it possible to design such information systems. Today, this technology is still not widespread enough, but in a very short time, the largest part of the SME-sector will be able to use it. It is reasonable, therefore, to rely on computer network technology to a larger extent in designing information systems for innovation support.

4.3 EXTERNAL RELATIONS IN THE INNOVATION PROCESS

4.3.1 Few SMEs have Innovation Partners, in Particular Other than Customers and Suppliers

Due to the fact that innovation is an interactive process both within a firm and between firms and other organizations, external relations are very important for the innovation process of firms (Kline and Rosenberg 1986; Dosi 1988; Malecki 1997) (see also Chapter 2). These relations often go beyond short-term market transactions and include more durable trust-based partnerships (DeBresson and Walker 1991; Cooke and Morgan 1993) and 'untraded interdependencies' (Storper 1997). Usually, there are various kinds of actors and organizations involved in the innovation process which interact in innovation systems (Lundvall 1992; Nelson 1993; Edquist 1997): customers, suppliers, competitors, service firms, universities and research organizations, technology centres and transfer organizations, finance and training organizations to name the most important. SMEs involved in innovation networks can be innovative also without strong internal R&D activities. They can rely on tacit knowledge, benefit from complementarities in local networks and from common learning, and rely on local institutions and resources (Cooke and Morgan 1998; Asheim and Cooke 1999). These are only potential benefits, however. Whether they can be realized depends on the networking capabilities of SMEs and these seem to be rather limited. The literature review in Chapter 1 shows that SMEs face many innovation barriers and have a low capacity and inclination for networking. According to the Community Innovation Survey (Eurostat) and a survey of regional innovation systems in Europe (Cooke et al. 2000) SMEs seem to be not only less innovative than larger firms but also less often engaged in networks.

What are the results of the SMEPOL project regarding the innovation networks of SMEs? The overall finding is that SMEs rarely have external relations in the process of innovation, clearly less than large firms. If there are external relations, then they are usually within the value chain. Of course, certain types of relations to external actors like arm's-length relationships do exist in the case of many firms, but they can hardly be called 'interactive'. Co-operative innovation projects, characterized by intensive collaboration and information exchange according to shared objectives, are very rare. It is a particularly serious deficit that the interaction with knowledge providers, both from science and technology, is very limited. Information and knowledge tends to be restricted to the well-known market leading to dependency either on strong business partners, usually dominant customers, or small markets for specialized products or services without being able to substitute if this market crumbles. Table 4.2 presents a brief overview of the SMEs' external relations in their process of innovation for each of the study regions.

That small firms are less willing to co-operate than large firms is confirmed by most regional studies. It is a widespread problem that SMEs are not willing or able to co-operate in their innovation process. In Upper Austria large firms are particularly more often engaged in innovation cooperation outside the region, in Austria as well as abroad, and with partners from science and technology. Also in Wallonia and Norway SMEs are less frequently interacting with external innovation partners than large firms. In Norway co-operation practices are more rarely in firms with less than 50 employees. The Italian investigation gives some evidence that willingness and ability to co-operate correlate to some extent with the state of the regions' economic development. In general, external relations in the innovation process are far more frequent in well-developed Lombardy than in the less developed and institutionally 'thin' region of Apulia. Firms in Apulia are as a consequence hardly willing or able to interact with their environment. This strong inward orientation of companies applies more or less to all types of potential innovation partners except for suppliers.

The predominant role of customers and suppliers over knowledge providers like universities and intermediaries like technology centres can be found in most regional studies. In Upper Austria, Limburg, Denmark, Norway, and Lombardy SMEs are hardly co-operating with universities, research organizations, and consultants. In particular the Danish stainless-steel cluster seems to be extremely customer-oriented. The Norwegian study shows that partners from science are more often used by large firms and R&D-intensive firms. Lower-technology and mature industries like boat building and fish processing have almost no links to science. There, collaboration concentrates on the local production system, but there are also barriers for innovation partnerships especially between yards and suppliers.

The technological level of a firm is very often the decisive characteristic distinguishing between different levels of willingness or ability to co-operate. This applies primarily to scientific partners, but to some extent also to busi-

Table 4.2   Typical patterns of external relations in the innovation process of SMEs

External relations to	Importance for SMEs in Denmark
Customers/suppliers in the region:	This was the most important type of interaction for most SMEs. Regional suppliers were also shown to be of relevance for large firms.
Customers/suppliers outside the region:	Within the relatively small territory of the Triangle Region, more than 20 per cent of new customers were located in that area. It indicates a widespread cluster of interrelated SME suppliers linked to a few corporate players. Further, when looking at both the Triangle Region and the rest of Denmark, in terms of 'new market' and 'new type of customer', the latter category was always significantly more important. Conversely, new markets are much more likely to be developed outside of the region.
Technology centres in the region:	In general, of some importance for firms.
Consultants/services:	In general, of little importance for firms.
Universities/research org. in the region:	In general, of little importance for firms.
Universities/research org. outside the region:	As far as the food-processing industry is concerned, the level of co-specialized research and research-based education was relatively low. There was a very slight preference to collaborate with foreign enterprises and institutes, rather than those available in Denmark.
Regional public support institutions:	Generally not very important.
National public support institutions:	Approximately one third of the responding firms made use of these institutions.
External relations to	Importance for SMEs in Limburg
Customers/suppliers in the region:	The most important type of interaction for most SMEs. Regional suppliers, but not regional customers, are also relevant for large firms.
Customers/suppliers outside the region:	The most important type of interaction for large firms, less frequent in the case of SMEs, especially outside the neighbouring regions. For certain firms, however, one or two international partners may be very important. More innovative SMEs are also more often engaged in innovation partnerships with customers and suppliers and on a larger geographical scale.

Table 4.2 (continued)

External relations to	Importance for SMEs in Limburg (continued)
Technology centres in the region:	They have few relations to SMEs, slightly more to large firms. SMEs that use them are more innovative in terms of resources as well as output than the ones that do not.
Consultants/services:	Far less important for SMEs than customers or suppliers. More frequently used than universities/research organizations and technology centres.
Universities/research org. in the region:	Of very little importance for firms in general, especially manufacturing firms. Nevertheless, there are some relations between SMEs and R&D units of large firms, not only via buyer–supplier relations (e.g. KIC), but also where R&D units do research for third parties (e.g. DSM and the Research-voucher project).
Universities/research org. outside the region:	More important than regional science organizations, but primarily for large firms. Especially in Northern Limburg SMEs have relations with universities in neighbouring regions which have technical faculties like Eindhoven; for example in the form of trainee agreements. International relations are rare. If there are such relations, then only with universities located in the close neighbourhood, in Flanders, Wallonia (Liège) and Germany (Aachen).
Regional public support institutions:	In terms of frequency they are not very important. There is no difference between technology centres and innovation-support institutions in this respect.
National public support institutions:	Not very frequently used, more by large firms than SMEs and more by innovative firms than less-innovative firms. Regional institutions are predominant in Limburg in delivering European, national and regional-specific innovation support to SMEs. Large and more innovative firms operate in a national and European system of innovation and innovation support.

External relations to	Importance for SMEs in Lombardy and Apulia
Customers/suppliers in the region:	Customers and suppliers in general are an important source of information for SMEs' innovations. Suppliers are mostly located in the regional area in Lombardy and in Apulia as well. Proper forms of co-operation among firms for innovation purposes are rare in both case studies.

Table 4.2 (continued)

External relations to	*Importance for SMEs in Lombardy and Apulia* (continued)
Customers/suppliers outside the region:	Suppliers outside the region are less important than those located within the region.
Technology centres in the region:	In both study regions over half of the investigated firms have made contact with regional technology centres.
Consultants/services:	As far as the development of new products is concerned, SMEs ask for external assistance quite frequently in Lombardy. They are very important, especially in Apulia, in giving assistance to firms applying for innovation support.
Universities/research org. in the region:	The university is, in general, not a usual partner of SMEs. In the case of Lombardy, SMEs contact mainly departments and organizations located in the regional area.
Universities/research org. outside the region:	In Lombardy universities/research organizations located there are more important than those outside the region.
Regional public support institutions:	The contacts with public institutions are in general on a bureaucratic basis. At regional level CESTEC, which is specifically constituted for supporting SMEs in Lombardy, is evaluating the eligibility of the applications to regional innovation-support laws. Little space is given to the development of innovation projects.
National public support institutions:	Not relevant.

External relations to	*Importance for SMEs in London*
Customers/suppliers in the region:	The majority of SMEs are supplying regional markets, especially in food and business services. Suppliers are less likely to be regional.
Customers/suppliers outside the region:	Larger SMEs are more likely to be involved in national and international markets, particularly firms in engineering.
Technology centres in the region:	There are no regional technology centres as such, but those technology-support centres that exist are little used (according to the control group survey).
Consultants/services:	Most important source of external assistance, particularly in engineering sectors. The majority of firms meet their needs via the market, particularly with respect to technical support.

Table 4.2 (continued)

External relations to	Importance for SMEs in London (continued)
Universities/research org. in the region:	Little used overall. Those few firms that do use them are mainly high-tech. HEIs used are not necessarily within the region.
Universities/research org. outside the region:	Some use of universities and HEIs by a small minority of firms in high-technology sectors.
Regional public support institutions:	There have been a number of EU-supported projects in the ELLV Ob. 2 area, including a BIC that is contributing to innovative new venture creation. It is difficult to assess the overall impact, but some of these projects have found it difficult to stimulate interest from SMEs.
National public support institutions:	Business Link (BL) is quite well-known and more frequently used than other public sector providers. BL is a national initiative organized on a sub-regional basis and provides an integrated approach to business support.

External relations to	Importance for SMEs in Norway (a) Non-R&D-intensive and regionally isolated SMEs in Southern Norway (b) SMEs in the fish-processing industry in Northern Norway
Customers/suppliers in the region:	(a) Regional customers are important for some SMEs serving the local market. However, the local market is generally not very demanding. Regional suppliers are of almost no importance. (b) Of very little importance, as customers and suppliers are mainly national and international.
Customers/suppliers outside the region:	(a) Demanding national/international customers are important for some local subcontractors. However, lack of systematic feedback from customers is a weak part of the innovation process in SMEs serving the final market. Contact with national/international equipment suppliers are important for process innovations in firms. Some component suppliers are also important. (b) Customers are the most important sources of information for the fish-processing industry. Suppliers of equipment and materials are important for process innovations.

Table 4.2 (continued)

External relations to	*Importance for SMEs in Norway* (continued)
Technology centres in the region:	(a) Of no importance for SMEs. (b) There are no regional technology centres which are relevant for the fish-processing industry.
Consultants/services:	(a) Of generally less importance than customers and suppliers, but more frequent collaborators than universities and research institutions. (b) Of very little importance, even less than research institutions.
Universities/research org. in the region:	(a) Very few SMEs have contact with regional research institutions. The few examples are initiated by the research institutions and triggered by policy instruments (like RUSH). (b) Of some importance, as Northern Norway has a university and research organizations specialized in the fishing industry. However, few firms see these organizations as important sources of innovation.
Universities/research org. outside the region:	(a) Of almost no importance, because only few SMEs of this kind use R&D-based knowledge in their innovation process. (b) Of little importance. However, the regional institutions (see above) are actually national organizations located in Northern Norway.
Regional public support institutions:	(a) Very important for co-financing innovation projects in firms. (b) Very important as a financier of innovation projects.
National public support institutions:	(a) Of little importance. Although the money to support innovation projects in firms comes from national funds, they are delivered by regional organizations. (b) Of some importance through branch-specific research programmes.

External relations to	*Importance for SMEs in Norway* *– More tech. advanced SMEs in reg. clusters in Southern Norway*
Customers/suppliers in the region:	Formal and informal contact with local customers and users has historically been highly important, especially for incremental innovations and in some clusters. It is still of some importance. Regional suppliers are important in some clusters. Their importance is increasing, as more long-term and binding co-operation, also including some innovation activity, between customers and suppliers evolves.

Table 4.2 (continued)

External relations to	Importance for SMEs in Norway (continued)
Customers/suppliers outside the region:	Large, mainly national customers often act as early and demanding customers, and are very important for innovation activities. Some specialized suppliers are found outside the regional cluster.
Technology centres in the region:	Of nearly no importance, except in one cluster – a technology centre in Jæren – where it has been historically of great importance for process innovations.
Consultants/services:	In some cases specialized consultants are highly important as designers in the ship industry.
Universities/research org. in the region:	Generally of little importance, as they cannot fulfil the requirements in technology-based SMEs which need very specialized competence.
Universities/research org. outside the region:	Of large importance in many technology-based SMEs. Their importance is still increasing, as firms upgrade their innovation activities using more R&D-based competence.
Regional public support institutions:	Also highly important in technology-based SMEs.
National public support institutions:	Especially important for technology-based SMEs, as these enter R&D projects co-financed by the Norwegian Research Council.

External relations to	Importance for SMEs in Upper Austria
Customers/suppliers in the region:	The most important type of interaction for most SMEs. Regional suppliers are also relevant for large firms.
Customers/suppliers outside the region:	The most important type of interaction for large firms. It is less frequent in the case of SMEs, in particular with partners outside Austria. SMEs which are more innovative, having introduced products new to the market or newly developed technologies, are also more often engaged in innovation partnerships with customers and suppliers.
Technology centres in the region:	Centres have very few relations to SMEs, slightly more to large firms. Especially SMEs with at least 10 employees and belonging to traditional industries such as wood and furniture and metal products do not use their services.
Consultants/services:	Less important for SMEs than customers or suppliers, but still more frequent than universities, research organizations and technology centres.
Universities/research org. in the region:	Of little importance for firms in general.

Table 4.2 (continued)

External relations to	Importance for SMEs in Upper Austria (continued)
Universities/research org. outside the region:	More important than regional science organizations, but primarily for large firms. Within the SME sector the most active users of science partners are firms in electronics and (advanced) plastic products, firms which have developed new technologies on their own, and firms where innovative products account for an above-average share in sales.
Regional public support institutions:	In general, not very important.
National public support institutions:	Frequently used, more by large firms than SMEs. The importance for the SME sector varies between certain industries, being most important for the metal industry and services. The national is the predominant level of Austria's innovation support system.

External relations to	Importance for SMEs in Valencia
Customers/suppliers in the region:	For the final producers of the sectors studied, the most important interaction partners are suppliers within the area. For the subcontractors of these sectors, the most important partners are the customers within the area. In both cases the frequency of interacting firms is very high (90%).
Customers/suppliers outside the region:	Less important than regional customers and suppliers. But still 40% of SMEs have contacts with national and international suppliers and customers.
Technology centres in the region:	They are very important partners in the innovation process. 60% of SMEs have contacts with the investigated technological institutes.
Consultants/services:	Less important than technology centres. Only 26% of the SMEs indicated links to them.
Universities/research org. in the region:	In general, very little importance for the firms.
Universities/research org. outside the region:	In general, very little importance for the firms.
Regional public support institutions:	The most important level of public support is regional. It is important for 40% of the SMEs (IMPIVA organization).
National public support institutions:	Less important than the regional institutions (only 30% of SMEs).

Table 4.2 (continued)

External relations to	Importance for SMEs in Wallonia
Customers/suppliers in the region:	Very important type of interaction for the smaller independent SMEs. The average Walloon SME finds innovative ideas either within the firm or through market contacts, most often clients and customers.
Customers/suppliers outside the region:	The most important partner for a large number of SMEs with international markets. The rate of openness of firms is very high in Wallonia according to European standards. The category of 'intra-group' relationships, i.e. relationships with clients or customer firms that belong to the same multinational firm, is also very relevant for a large group of foreign-owned SMEs, and for larger firms.
Technology centres in the region:	Few relations to SMEs, slightly more to large firms.
Consultants/services:	Private consultants have a very limited role as innovation partners for Walloon SMEs. Larger firms use them more, but the nature of the services bought is not clear.
Universities/research org. in the region:	Not distinguished, see next category.
Universities/research org. outside the region:	Universities are co-operation partners for 20% of the independent SMEs. This share is higher for SMEs which are beneficiaries of public support, showing the orientation of policies towards the linkage between research and industry. Large firms are well linked with universities, but the CIS2 inquiry does not clarify whether these links are with national or international universities. Qualitative results from the SMEPOL inquiry show that relationships with EU universities are quite frequent, as the main reasons for co-operating rest on excellence rather than proximity. The nature of co-operation varies according to the degree of maturity of the technology used in the firms: testing and procedures relating to norms and standards are more frequent forms of collaboration than joint research.
Regional public support institutions:	Regional support is relevant only for larger firms, not for SMEs.
National public support institutions:	This type of partner is not relevant any more in the federal context of Belgium, where regions have autonomy of power in the areas related to innovation support.

ness partners. In Upper Austria the most active SMEs regarding innovation co-operation are higher-technology and more innovative firms, mainly in electronics. Nevertheless, the general level of co-operation activity in these cases is still low. In Norway firms performing R&D are more willing or able to co-operate. This is also confirmed by the London study. It was found that SMEs of higher-technology industries such as computer and machinery equipment manufacturing and electrical and instrument engineering tend to use external advice and consultancy for product innovation more often than food processing and business services. In Valencia universities are, in general, of little importance. But the ceramics sector frequently uses universities and public research institutes to collect innovation-related information.

Exceptions concerning the negligible role of innovation partners outside the value chain are consultants in London and technology centres in Valencia. The London case is especially remarkable, where private sector consultants belong to the most important innovation partners of SMEs. They are significantly more often used than the support scheme, 'Business Link', suppliers, and higher education institutions. The dominant role of clients is confirmed by the Valencia study too; nevertheless technological institutes are remarkably important, surpassing suppliers in stimulating innovation and only slightly less important than suppliers as sources of information. The case of Lombardy shows that it is particularly the smallest size class of SMEs with not more than 20 employees that needs services from technology centres most.

The case of Limburg shows that business and professional associations, regional as well as national, can be important mediators for SMEs to find innovation partners from the firm sector in addition to their own buyer–supplier relations. In Limburg these are mostly informal cross-sector associations, like a kind of Rotary for regional or local entrepreneurs. SMEs can also find partners and respective information and services from the same sector at national professional associations, like the Dutch association for furniture manufacturers. In Wallonia the professional associations are especially relevant in the traditional sectors such as steel fabrications. There they are well-known to firms and can play a stimulating role in some cases. They also act as deliverers of policy instruments on occasions. Nevertheless, the real co-operation goes through the collective research centres, to which these associations are usually closely linked.

A further deficit regarding innovation partners is the low level of inter-activity especially with business partners. Often these innovation relations cannot be qualified as 'interactive'. For example, clients and customers seem to be rather important partners in Wallonia, but obviously more as sources of innovation-related information than as closely interacting co-operation partners. SMEs especially are hardly co-operating in their innovation activities. If there are co-operations then those occur most frequently with universities, rather than with customers, suppliers, and consultants. Nevertheless, these cases are very rare. There seems to be little variance between types of Walloon SMEs in this respect. As far as interactions with universities are concerned, it has to be considered, however, that lower technology firms have more service-type relations with science, such as testing. Real co-operations like joint research projects can only be found in some cases of higher-technology firms.

Regarding policy this implies an obvious need to support the capacity or ability of SMEs to co-operate. For this purpose three requirements have to be fulfilled. First, there have to be sufficient and adequate knowledge-providing functions in the region or the country. Such functions are lacking especially below the level of science and beyond technology, for instance management and finance. Second, interactions with existing potential providers of inputs to the innovation activities of SMEs have to be mediated and stimulated. Networking with all kinds of partners from business, science and technology within the region is very important for SMEs to overcome the limits in scale and scope of their innovation capacity. Third, due to the fact that many SMEs are not aware of the advantages of co-operation or are unwilling to co-operate, the value of innovation co-operation has to be demonstrated, for example through information dissemination of successful co-operation projects. It is also necessary to raise awareness of the problems resulting from a lack of interaction. Overall, a proactive policy approach and new ways of communication and consulting are required in order to overcome SMEs' barriers to networking.

4.3.2 SMEs are More Region-centred than Large Firms

As already argued in Chapter 2, several studies show the importance of the regional level in innovation networks (Autio 1998; Braczyk et al. 1998; Malecki and Oinas 1999; Cooke et al. 2000). One of the most important reasons is that proximity favours frequent face-to-face contacts. Proximity is particularly important regarding the exchange of tacit knowledge, a crucial resource for most innovation activities (Lundvall and Borrás 1997). According to Cooke et al. (2000) SMEs clearly concentrate their innovation relations more on their home region than larger firms. Overall, this has been confirmed by the SMEPOL case studies.

In Upper Austria and Limburg the primary spatial orientation of SMEs concerns the region. But there is not only a difference between SMEs and large firms in this respect, since also within the SME sector the region becomes more important with smaller firm size (Upper Austria, London, Lombardy). A too dominant focus on the region limits the scope of available technical information, technologies, and accessible markets, however (Camagni 1991; Cappellin and Steiner 2002). In Upper Austria this seems to be more pronounced in the case of traditional industries and the service sector. Of course, there are differences between SMEs, some involved in local production systems, as in the case of the Norwegian boat industry, footwear firms in Valencia and the stainless-steel sector in Denmark, and others regionally isolated interacting on national, sometimes even international levels. In London, for example, computer and machinery equipment manufacturing, electrical and instrument engineering are less dependent on regional markets than food processing and business services.

This raises the problem of a lack of adequate partners to co-operate with in the innovation process due to the limited scope of the region. This is usually one of the most frequent reasons for SMEs not to co-operate. In Upper Austria the 'lack of adequate partners' ranks second only to 'no need' as a co-operation barrier. But it can also be the other way round. In Limburg SMEs often lack the qualification as system suppliers for the metal and electrical industry and, therefore, cannot become innovation partners.

The general correlation between extra-regional markets and firm size should not conceal the existence of highly specialized engineering firms which are often small as well as export-oriented as the case of London shows. The problems they are facing concerning their export activities are especially serious for such firms because of the limited internal resource base.

In Italy there is – due to a pronounced core–periphery structure – a big difference in the importance of the region as a source of technical know-how for innovating firms. In Lombardy 61 per cent of firms indicated adequate sources in the local area, in Apulia the share is only 13 per cent. In Lombardy 39 per cent of firms perceived themselves being a part of a local innovation system, in Apulia only 6 per cent. Obviously, the 'thinness' of the regional innovation system (see Chapter 3) plays an important role here.

Support instruments, which focus on the mediation of market and technology information, have also to engage in making extra-regional information accessible to SMEs. Regional networks are a necessary but not sufficient condition for stimulating innovation. To focus on the regional information and interaction space only cuts the SMEs off from a wide range of potential sources of information and innovation partners. It is especially important for peripheral and institutionally 'thin' regions to maintain or establish links beyond the region. Technology centres might perform the central role as an information gatekeeper channelling communication with extra-regional sources of information. They are easily accessible for local SMEs and can therefore act as the primary mediator between the regional SME sector and a wide range of knowledge providers and potential innovation partners, business as well as science and technology, outside the region. This function, however, can be performed only if the centres have enough adequately skilled professional staff.

4.3.3 Differences in the Regional Institutional Settings Lead to Different Preferable Entry Points for SME Support

Examples are the particular roles of technology centres in Valencia and private sector consultants and services in London. Important functions to support innovation might be missing in the overall institutional setting of a region. In some cases there is already an institutional framework established which makes it easier to add a missing function. In Upper Austria, for example, this applies to technology transfer and consultancy which is hardly performed by technology centres and a rather small university. In other cases there is not even an adequately developed institutional basis. This seems to apply, for example, to external innovation-related services in Apulia.

The implications for policy are that, in reorganizing the innovation support infrastructure, the focus should be on the already existing elements of a support system. There are no standard blueprint solutions which are adequate to any region. This means that, first, weaknesses in the institutional setting have to be removed, if they are important and not taken over by other institutions. Only in the case that certain institutions are missing at all or cannot be improved at reasonable expense, new institutions have to be established. Therefore the first step would be that support organizations become more proactive. In the case of incubation-focused technology centres this often concerns the neglect of relations to external SMEs. In the case of direct support a frequent problem is the closed clientele. Elements exist, but their reach has to be extended. If this is not sufficient, then the missing functions have to be added to existing institutions, broadening their range of support activities. The establishment of new institutions should be the last step, however, in order to avoid the proliferation of institutions. Any new functions or institutions have to be integrated in a coherent and complementary support system which requires continuous monitoring of the effects of the support instruments (see also Chapter 8) and, if necessary, adequate adjustments. In this context, it is necessary not to concentrate on public support activities only, as already put forth in Chapter 1. Attention must be paid also to the complementarity of public and private providers of support. Public support should not compete with private services but concentrate on problems caused by market failure. Nevertheless, public activities will remain important, because many services are not profitable and private consultants focus on the most profitable services. As a consequence, more latent deficits and long-term structural weaknesses often remain untouched.

4.4 PROBLEMS CONSTRAINING INNOVATION

The problems constraining or preventing innovation are very diverse depending on type of region and of SME. Nevertheless, some results apply to most SMEs in general (see Table 4.3). The analysis of barriers to innovation is further complicated by the fact that firms may be unable to consider all their problems. Some may be recognized, some might even be over-assessed, while others are under-assessed, and some might be overlooked completely.

Problem patterns of lower-technology firms are usually different from higher-technology firms and, in general, more serious. According to the Walloon results, technological barriers are more serious in lower-technology, traditional sectors. In Norway lower-tech firms like boat builders have technological deficits because they rely on practical experience sufficient for incremental, often imitative, change but insufficient for more advanced innovations. The required technical know-how is certainly below research level and often adaptive in character. Technologies are available elsewhere and adjusted to context of boat-building.

It is important not to neglect other types of problems than technical or financial which are constraining innovation. Often firms do not sufficiently recognize them. It is especially the value of co-operation for innovativeness which is under-assessed by many SMEs. In Upper Austria the most frequent reason for SMEs not to interact is 'no need'. In addition, market research activities are rarely performed. In Apulia some firms indicated problems such as a lack of market information, difficulties in co-operating with research centres, a lack of external services, and the geographical or cultural distance from sources of innovations as seriously constraining their innovation activities. Not surprisingly, in Apulia, the less innovative Italian region, more firms indicated barriers than in the more innovative region, Lombardy. But even here, the majority of firms did not consider market access or information as a relevant problem category.

Of central importance are strategic deficits and organizational weaknesses of SMEs. A frequent strategy deficit in the case of SMEs is the narrow customer focus making their innovation process dependent on their clients (Upper Austria, technologically advanced firms in Norway, Limburg, Denmark). But there are also cases where firms lack feedback from their clients (low-tech, non-R&D-intensive firms in mature industries in Norway).

Table 4.3   Types of problems constraining innovation in SMEs

Problem categories	Relevance for and awareness in SMEs in Denmark
Finance/risk:	Costs in terms of employee time, investment in new equipment, and monetary input were cited as problems of some importance, more for product innovation than for process innovation. The financing of development costs was also cited as an influencing factor, again, having more impact on the product side.
Personnel/ qualification:	The most problematic external barrier was found to be a lack of qualified staff, especially in the fields of electronic engineering, software computing and, more generally, qualifications in electronics. There were less problems in the low-tech stainless-steel sector. When examining recruitment of qualified personnel, firms have some problems. The problem is marginally worse on the process-innovation side.
Technology/technical know-how:	In line with recruitment problems, firms can lack the necessary know-how, especially in the case of product innovation.
Market access/ information:	Some firms found that lack of knowledge of the market (in the case of product development) or technology (in the case of process development) was a barrier. The problem was more serious in the case of product development.
Time/organization:	Time was cited as the second strongest barrier for product innovation and the third for process innovation.
Strategy:	As part of the overall strategy, there is an implicit preference to form relationships with customers as opposed to suppliers, even though firms feel that there is a lack of suitable customers to collaborate with. Internal barriers play a strategic role when it comes to the pattern of innovation, and the innovation of new products is largely driven by the need to 'keep customers'. Subsidy schemes and new regulations play virtually no role in influencing strategic behaviour. Investment in competence building in 'management' is very low, lower than in other areas of business administration. It is a paradox then, that 'management' contributions are perceived by firms to be very important for both product development and process development.

Table 4.3 (continued)

Problem categories	Relevance for and awareness in SMEs in Limburg
Finance/risk:	This is an obvious and often-mentioned barrier, but it is not the most important one. Mainly non-innovating firms mention finance as a barrier or an excuse not to engage in innovation. Firms which are aware of the importance of innovation and which are engaged in it do not regard finance as the main barrier they had to overcome.
Personnel/ qualification:	More important to smaller and less innovative firms and in specific technological fields. Shortages on the labour market have recently become a major problem, not only for innovation and not only for small firms.
Technology/technical know-how:	Unavailable technology, either missing or too expensive, is a relevant problem.
Market access/ information:	This is a frequent problem for SMEs, especially regarding customer and supplier dependency. Firms mainly have a reactive and only rarely a proactive attitude towards in- and output markets.
Time/organization:	The time problem is widely recognized, especially by the SMEs that do not innovate. Organizational deficits, on the contrary, are less often considered. The innovating firms are aware of the organizational and management deficits.
Strategy:	There is little persistence and commitment to enter new markets and to diversify. Introverted innovation activities, little search for information beyond the value chain, and general reluctance to use external advice and risk capital are major strategic weaknesses.

Problem categories	Relevance for and awareness in SMEs in Lombardy, Apulia
Finance/risk:	The lack of finance is a frequently indicated problem, especially by smaller SMEs with less than 20 employees. The high costs related to the development of innovative projects are a great concern for SMEs. This obstacle is more crucial in the case of Apulia than in the case of Lombardy since a much higher percentage of the investigated firms in Lombardy would have made the investments for the innovative project also without support.
Personnel/ qualification:	The lack of qualified personnel is perceived as an important obstacle to innovation in SMEs. Often insufficient steps are taken to face this problem in the firm's strategy.

Table 4.3 (continued)

Problem categories	Relevance for and awareness in SMEs in Lombardy, Apulia (continued)
Technology/technical know-how:	It is not considered as an obstacle by most of the investigated firms. The production is often highly specialized, so the process of learning by doing plays a dominant role in determining the firm's success. Experience, however, is not sufficient, when new technologies are introduced.
Market access/ information:	It is not recognized as a problem by the large majority of the investigated SMEs.
Time/organization:	Many investigated SMEs face organizational problems taking steps to introduce quality certification.
Strategy:	In many firms there is a lack of awareness of the strategic role of the innovation activities they are carrying out.

Problem categories	Relevance for and awareness in SMEs in London
Finance/risk:	Finance is the most commonly identified barrier (49% of firms with respect to product innovation), although most firms do not actively seek external finance.
Personnel/ qualification:	Shortages of skilled labour were identified by 10% of firms as a barrier to new product innovation and to process innovation. This applies particularly to engineering firms.
Technology/technical know-how:	The majority of firms recognized the need to raise the firm's level of technology and technology competence. This applies particularly to the engineering sector.
Market access/ information:	This was not explicitly identified as a problem by our control-group survey. Two-thirds of surveyed firms had introduced some new marketing methods during 1993–98. Of the control-group firms, 67% had developed new markets of some type during 1993–98. However, our surveys of SMART winners and LVBIC clients confirm that the cost of marketing is a significant barrier for innovative small firms, particularly new firms.
Time/organization:	The time problem was highlighted by a minority of respondents (20% in the case of process innovation). Organizational issues were not explicitly identified in our study.
Strategy:	There is an apparently high propensity to enter new markets: 67% of control group firms had developed new markets of some type during 1993–98.

Table 4.3 (continued)

Problem categories	Relevance for and awareness in SMEs in Norway *(a) Non-R&D-intensive and regionally-isolated SMEs in Southern Norway, (b) SMEs in the fish-processing industry in Northern Norway*
Finance/risk:	(a) This problem was indicated by several firms, in particular concerning capital to be invested in long-term projects which do not promise rapid return. (b) Excessive perceived economic risks are seen as important factors hampering innovation by fish-processing firms.
Personnel/ qualification:	(a) Recruitment of qualified personnel to carry out innovation projects creates an important bottleneck for a majority of firms. Firms are aware of this problem. (b) There are serious problems in recruiting and maintaining stable labour relationships, also caused by the out-migration from fishing communities and unstable working conditions in periods of resource shortage. There is awareness of this problem in the firms.
Technology/technical know-how:	(a) Firms often have good experience-based know-how, but may lack formal know-how, necessary for more systematic innovation activity. Firms may not be fully aware of this constraint as they often only rely on step-by-step improvements. (b) Lack of technical information is indicated by firms as restricting innovation, connected to a low formal competence in the workforce.
Market access/ information:	(a) There is a lack of systematic feedback from the market in several firms. (b) Firms often lack feedback from the final market, as they generally produce 'half-finished' products.
Time/organization:	(a) How to free up personnel in the firm and make them available for innovation activity is an important constraining factor for several firms that firms seem to be aware of. (b) Organizational rigidities are mentioned as the most important factor restricting the innovation process. There is often too little flexibility to meet changes in the resource situation and the market.
Strategy:	(a) In some cases entrepreneurs do not want their firm to grow too much in scale and employment, which also restricts innovation activity.

Table 4.3 (continued)

Problem categories	Relevance for and awareness in SMEs in Norway (continued)
	(b) Maybe too much emphasis on process innovation and cutting costs and too little on product innovation and market research. There is a lack of co-operation with other local fish-processing firms and the regional R&D milieu.

Problem categories	Relevance for and awareness in SMEs in Norway *– More tech. advanced SMEs in reg. clusters in Southern Norway*
Finance/risk:	This seems to be less of a problem in technology-based SMEs, because they often have easier access to external funding.
Personnel/ qualification:	Problems in recruiting highly qualified and experienced workers were frequently indicated by firms. However, location in a regional cluster may facilitate the recruitment.
Technology/technical know-how:	Generally, the firms have adequate internal technological know-how. However, they indicate a lack of relevant competence in regional R&D institutions and a fear of stagnation in some national R&D institutions as problems. There is a lack of potential collaborators in technological innovation projects.
Market access/ information:	Technology-based firms have generally close co-operations with some large and demanding customers.
Time/organization:	Firms frequently indicate little time left for long-term work on innovation projects during busy periods, as key personnel are busy carrying out the daily work.
Strategy:	In general, innovation forms an important part of the firms' strategies. The lack of close co-operation with local suppliers and other firms may hamper innovation activity.

Problem categories	Relevance for and awareness in SMEs in Upper Austria
Finance/risk:	This problem was frequently indicated, especially by the smallest SMEs with less than 10 employees, producer services, and capital-intensive manufacturing like machinery. Firms are fully aware of this problem, sometimes even over-assess it regarding available sources of funds.

Table 4.3 (continued)

Problem categories	Relevance for and awareness in SMEs in Upper Austria (continued)
Personnel/ qualification:	More important in higher-technology sectors, producer services, and less innovative firms without innovations which are new to the market.
Technology/technical know-how:	Unavailable technology, either missing or too expensive, is a relevant problem in some industries, but frequently of low importance. Even less important are deficits in technical know-how except for higher-technology sectors. Within the firms' traditional markets, deficits seem to be recognized, but know-how required to enter developments beyond the traditional market is rarely considered.
Market access/ information:	Deficits in this respect are often under-assessed by SMEs, especially regarding customer dependency, which is most serious in the case of the more innovative firms. Deficits in marketing are more frequently recognized, but here this is more serious in the case of the less innovative firms.
Time/organization:	Time problem is widely recognized. Organizational deficits, on the contrary, are less often considered.
Strategy:	There is little willingness to enter new markets and to diversify. In many firms introverted innovation activities, little search for information beyond the value chain, and general reluctance to use external advice and risk capital are strategic weaknesses.

Problem categories	Relevance for and awareness in SMEs in Valencia
Finance/risk:	This is a widespread problem, 50% of the investigated SMEs mentioned it.
Personnel/ qualification:	This is also a relevant problem, but less important, 28% of the SMEs mentioned it. It was more frequently mentioned by the more innovative SMEs of the ceramic sector.
Technology/technical know-how:	In general, this is not a frequent problem, 13% of SMEs mentioned it. It is more important in the case of the more innovative SMEs
Market access/ information:	This problem is of negligible importance, only 9% of SMEs mentioned it, but it might be that few firms are aware of this deficit. It was more frequently mentioned by the more innovative SMEs.
Time/organization:	Not mentioned by the firms.
Strategy:	Not mentioned by the firms.

Table 4.3 (continued)

Problem categories	Relevance for and awareness in SMEs in Wallonia
Finance/risk:	The majority of innovative SMEs report a lack of appropriate sources of funding as a major barrier to innovation, followed closely by the problem of high costs to innovation and by the excessive economic risk perceived to innovate. Financial barriers are also the main barriers for larger firms, but they seem to hinder the innovation process less frequently.
Personnel/ qualification:	Lack of qualified staff is an important barrier to innovation for more than one-third of SMEs. According to the interviews, this problem seems to concern more the shortage of middle-level technicians than of higher qualifications.
Technology/technical know-how:	Lack of technological information is constraining innovation only in a few cases. It is not a main barrier, certainly not for larger firms. Newly innovating SMEs of traditional sectors express the need to access research equipment for small-scale experiments.
Market access/ information:	Market information is not proposed as an important barrier for innovation in surveys. SMEPOL interviews have shown, however, that SMEs often envisage innovation from the point of view of improving processes or the use of new technology, but are not so well equipped to analyse market trends and opportunities.
Time/organization:	Interestingly and, perhaps, not unexpectedly, in the CIS2 survey, organizational rigidity is reported twice as frequently for larger firms than for smaller ones as a barrier for innovation. Actually, it is one of the main barriers for larger firms. Is it the expression of a more acute problem in larger firms or the reflection of their enhanced understanding of the problem? The 'time' constraint mentioned by many SMEs in interviews can be interpreted as a reflection of weaknesses in strategic thinking.
Strategy:	The SMEPOL inquiry showed that one explanation for the lack of innovation dynamics is the difficulty for SMEs, and especially micro-firms, to manage growth necessary for innovation. Combining product and process innovation and company development seems to be a major difficulty for Walloon SMEs.

In some regions this is reinforced by the neglect of systematic search activities concerning new market opportunities (Upper Austria, Limburg, Wallonia). In other regions, however, the willingness to enter new markets was more widespread (London).

The 'time' problem is usually better recognized. The problem of daily work overload of very few persons or even a single person in SMEs impeding or delaying innovation projects was frequently indicated in the investigations done in Upper Austria, Norway, Limburg, Wallonia, and Denmark, less in the case of London. The likely connection to organizational weaknesses seems to be less often recognized.

Small firms have typically serious growth-related problems restricting the potential to realize innovations. In London, for example, the financial problems and the need for external consultancy are strongest in the case of small firms between 10 and 49 employees and in the higher-technology sector. The size class 10 to 49 especially contains many firms intending to grow beyond the owner-centred management organization, which is a very difficult phase for firms presenting them with serious organizational and financial problems of growth.

Referring to the problem of awareness, this is not only a certain kind of problem which might be neglected by a firm. In many cases SMEs are not aware of possible or available solutions to problems. This can be due to a lock-in situation or the lack of capacity of key persons to deal with all relevant information. The example of London shows that even in a case where risk capital is offered, SMEs often resist using it for the following reasons:

- not convinced about getting funding
- too expensive (interest rates too high)
- minimize exposure to debt
- fear losing control when incorporating venture capital.

It can be seen that in targeting innovation support adequately, policy cannot rely on those problems which the firms are aware of, but that it is often necessary to first raise awareness of potential deficits not yet sufficiently recognized by many SMEs. It is often the insufficient strategic orientation of a firm that leads to neglecting certain aspects of innovation and the related potential deficits. Strategic weaknesses should become a relevant target of innovation support. This does not apply only to the actual process of innovation, but also to the exploitation and commercialization of innovations. On the other hand, problems should not be 'over-supported' and more resources should not be dedicated to solve a certain problem than are actually needed by the firms. This applies particularly to financial support, if it reaches primarily already innovative firms.

As a consequence, support has to aim at raising awareness of weaknesses which are not sufficiently recognized by SMEs, e.g. through proactive consultancy. Also, reliable information about the deficits and needs for innovation support has to be collected from the companies (see also Chapter 8). An interactive and regular information process on these issues might inform companies about typical innovation problems and effects of specific support instruments and it might in turn encourage them to contribute their own data and information in this respect. In such a process each firm, then, could compare its own situation with those of other companies and identify its respective position. This might raise awareness for problems so far neglected and it might stimulate innovation and reorganization.

The SMEPOL survey, furthermore, has shown that SMEs are often not aware of already available sources of innovation support and ways to access them. Therefore, companies should be informed extensively about such sources, helping them to solve specific problems, both offered within and outside the region and by private as well as by public providers. In addition, specific consulting regarding ways to access such support should be offered. Many SMEs lack the capabilities to make use of certain support instruments, even if they are interested in applying for them and belong to the respective target group.

4.5   CONCLUSIONS

Although SMEs' innovation activity and practice is quite heterogeneous, we also find common patterns. The studies conducted in the framework of the SMEPOL research project make clear that providers of innovation support have to find ways to deal with the following typical characteristics of SMEs' innovation activities:

- Innovation tends to be part of a defensive strategy focusing on market niches.
- Innovation takes place in close, sometimes dependent, relation with partners of the value chain, primarily customers.
- The willingness or ability to co-operate, especially with organizations outside the value chain, is very low.
- Market research is hardly performed.
- Research is rarely conducted in the process of product development which, however, does not prevent some firms from introducing advanced innovations.
- Human capital is of outstanding importance as resource for the innovation process.

- Lack of time seriously constrains the realization of innovation projects.
- The region is the dominant interaction and information space.
- Weaknesses regarding organization, strategy, market information and innovation co-operation are often not recognized as deficits constraining the ability to innovate.

As a consequence, an effective innovation policy for SMEs should basically try to:

- raise their awareness for innovation and networking, and
- enhance their capability in this respect by supporting not just R&D but also other activities such as market research, knowledge acquisition and management, and opening-up of new markets.

This requires in particular that we should:

- improve the firms' manpower and human resources for R&D and innovation,
- stimulate relations to partners also outside the value chain with universities, research and technology transfer organizations, and
- stimulate such networks not just within the region and the country, but also on European and, if necessary, global levels.

It is necessary, however, not to simplify the innovation processes of SMEs to a single typical pattern. A general finding of the studies is the heterogeneity of the SME sector regarding their innovation activities, the resources employed, the partners used, and the problems the firms are confronted with. Any innovation support of SMEs should, therefore, address the particular features and weaknesses of SMEs in the respective regions. It should be therefore both targeted with respect to types of SMEs and their problems and comprehensive regarding the innovation functions addressed. A certain degree of diversity as well as of flexibility of the support system will be required.

For policy this implies that effective innovation support requires reliable information about the deficits and needs for support. In this context it might be reasonable to establish a routine information process on support needs and support effects in firms. The procedure should be organized in an interactive way: direct and frequent contact with firms informs the providers about the expressed needs of firms and the effectiveness of their support activities. This is the basis for the adjustment of the support instruments. As an additional positive effect, informing the responding firms about the results of such surveys allows them to deduce their position regarding problems and needs relative to other firms, possibly raising awareness for problems so far neglected. The interactive design would probably raise the willingness of firms to participate in a repetitive survey, if the information returned to them is interesting and useful for them. Based on such information it would be possible to target innovation support for SMEs more precisely. Which are the main types of problems: finance and risk, personnel and qualification, technology and technical know-how, market access and information, time and organization, or strategy? In which types of firms are certain problems most serious? Are adequate solutions offered and accepted or ignored, or are they not even offered? Is it necessary to raise awareness regarding specific problems first, before offering support? Are there problems which seem to be 'over-supported'?

Support instruments have to be accessible for all levels of technology and size classes of companies, but can be particularly designed for industries which are of highest importance in a region. The support system should be designed for the whole range of SMEs and not just targeting high-tech firms. The primary objective of innovation support should be to improve the capacity to innovate, not to support the already innovative firms. Support instruments should concentrate on firms which need support most, and these are the less innovative or non-innovative firms. This requires another approach than project support, because there is an immanent focus on 'winners' in this kind of support. Project support is usually only accessible for firms which already have the suitable human capital to innovate but might lack funds to realize a certain development project. Other firms which lack the ability to innovate cannot benefit from such an instrument. This means that project-oriented support should be restricted to extraordinary innovation projects only and should not be the dominant form of innovation support.

Overall, the present chapter leads to the conclusion that the region, its innovation system, and its policy actors and organizations should play a central role in supporting SMEs in their innovation process. As was shown above, the region is in fact the dominant interaction and information space for many SMEs and it is also the most important source for human resources and skills. It is to be expected that actors and organizations from the region are generally better able to reach less innovative firms, find out their specific innovation problems, and raise the companies' awareness regarding innovation and related problems. Furthermore, the 'delivery mechanisms' of instruments are easier to organize in an interactive way at the regional than at national or European levels. This does not imply that national or European policy instruments for SMEs are becoming less important. What we would like to argue, however, is that support instruments should be co-ordinated at a regional level and be offered there in a coherent way in order to enhance their effectiveness (this aspect will be further analysed in Chapter 7). The demands for co-ordination and coherence make it necessary for the innovation support of SMEs to become a well-integrated element not only of the respective national but also the regional innovation system, as has been argued already by other observers (Braczyk et al. 1998; Malecki and Oinas 1999; Cooke et al. 2000).

PART III

EVALUATION OF INNOVATION POLICY INSTRUMENTS

5. Innovation policies for SMEs: an overview of policy instruments[3]

Gioacchino Garofoli and Bernard Musyck

5.1 INTRODUCTION

This chapter offers an overview of policy instruments identified within the SMEPOL study. A matrix (Table 5.1) is proposed, which takes into consideration the nature of the tools (finance, information, advice, and human resources) and the target groups to which these tools are aimed (individual firms or local and regional productive systems).

According to the findings of Chapter 4, the proposed typology reflects the main weaknesses of SMEs regarding innovation, as revealed in the SMEPOL studies of SMEs' innovation performance. These are (i) financial barriers, (ii) lack of accessibility to strategic information, (iii) lack of interaction with knowledge providers, and (iv) lack of adequate human resources.

The matrix takes into consideration the nature of the tools and the particular target groups to which the tools are aimed. We differentiate between two different types of target groups in the support schemes of technological transfer: firms (through schemes directly oriented to them) and the local and regional system (through the use of indirect schemes fostering external economies and the production of public goods).

The distribution of different innovation schemes in the matrix in Table 5.1 generates the following typology:

A. Firms-oriented support schemes:
A1. support schemes (through grants and loans) for innovation projects, the introduction of new products and R&D projects;
A2. introduction of research/technical personnel in SMEs.

B. System-oriented support schemes:
B1. policies based on technology centres and schemes fostering technological diffusion to SMEs;
B2. schemes fostering the role of innovation brokers;
B3. mobility schemes for researchers.

C.   Process-oriented support schemes:
C1.   proactive actions and initiatives;
C2.   upgrading of local and regional innovation systems.

The first group (A) comprises tools that are designed for individual firms, and recognizes financial barriers and the lack of technical competencies, as main barriers to innovation in SMEs. Until today, these schemes have been prevalent in many European regions. The central aim of these measures is the reduction of the cost of innovation faced by SMEs.

The second group (B) focuses on the indirect diffusion of technological competence through increased interaction between private and public actors. They include schemes in support of indirect technological diffusion for SMEs, schemes based on 'technology or innovation brokers' and schemes that promote the diffusion of technological expertise through the mobilization of researchers.

Table 5.1   Typology of evaluated policy tools

| Tools | Target groups | | |
	Firms-oriented (A)	System-oriented (B)	Process-oriented (C)
Finance	Support schemes for innovation projects, R&D projects (A1)		Proactive actions and initiatives (C1)
Information		Technological centres (B1)	Upgrading of local and regional innovation system (C2)
Advice		Innovation brokers (B2)	
Human resources	Technical personnel introduction schemes (A2)	Mobility schemes for researchers (B3)	

Categories A and B are relatively 'conventional'. What seems to be missing in most of the schemes under study is the existence of interactive mechanisms, which can foster learning processes, additionality, and the progressive networking of local actors and institutions. In short, this means the construction of a local and regional system of innovation. To this effect, a third category (C) is introduced which includes dynamic innovation schemes. This last category discusses interactive tools based on proactive actions and initiatives as well as new schemes aimed at upgrading the local and regional innovation systems as a whole.

This typology is a very simple one and is based on the main attributes of existing innovation policy schemes. It may help us to formulate critical reflections on the regional case studies, allowing us to propose policy lessons at the regional level.

In the next three sections we illustrate the above typology with the specific schemes introduced in the different study regions. We show the features of the various schemes, underlining the specific tools and measures relevant to the proposed typology. To conclude, some policy implications are discussed regarding the enhancement of the capacity of local and regional policy-makers to face the innovation challenge.

5.2   COMPANY-ORIENTED SUPPORT SCHEMES (A)

5.2.1   Support Schemes (through Grants and Loans) for Innovation Projects, the Introduction of New Products and R&D Projects (A1)

These innovation schemes and incentives are designed for individual firms, and underline the crucial role of the entrepreneur or the SME manager in the innovation process. These support schemes recognize the financial barrier as the main obstacle to innovation.

The schemes are rather traditional and are incompatible with the need to promote interactive learning processes. Moreover, they do not recognize the existence of local and regional innovation systems. Yet, until today, these schemes have been prevalent in many European regions. The central aim of such schemes is, as mentioned before, the reduction of the cost of innovation faced by SMEs. Through grants and loans, these schemes support the purchase of innovative and up-to-date equipment and also offer specific real services and consultancy.

It is useful to underline the difference between the schemes that foster the purchase of new equipment and machinery (A1.1) and the schemes that promote the introduction of new products and R&D investments (A1.2).

Support Schemes for Process Innovation

The following schemes foster the purchase of new equipment and machinery and are specifically geared towards individual firms: in Italy, national laws L. 1329/1965, L. 696/1983, L. 399/1987, L.317/1991, L. 488/1992, L. 598/1994; in Lombardy regional laws L.R. 34/1985 and L.R. 7/1993; in Apulia, regional schemes introduced mainly through structural funds. In Belgium, some schemes supporting material investments in Walloon development zones; and finally in Austria, the Regional Innovation Premium (RIP).

At the national level in Italy, incentive schemes are geared towards the introduction of new capital equipment by means of subsidized credit or capital grants. This is typically the case of national laws L. 1329/1965 (the so-called 'Sabatini Law'), L. 696/1983, L. 399/1987, L. 317/1991, L. 488/1992 and L. 598/1994, which support the use of new capital equipment. Some of these schemes (especially the 'Sabatini Law') have been relatively successful, due to simplified procedures.

The Regional Innovation Premium (RIP), in Austria, is a programme designed to support the economic recovery of old industrial regions, and to contribute to structural improvements and economic growth in peripheral regions. The programme, which is administrated by the ERP, offers non-repayable grants for investment projects with an innovation or technical content, as well as for the creation of new jobs. While size and sector do not play a role, only firms located in certain areas can benefit from the scheme. The amount of the grant depends on the impact of the companies' activity on the regional economy (relations with regional suppliers and impact on the level of qualifications). The scheme offers a clear SME focus. One of the weaknesses of the programme is its insufficient focus on producer services. Compared to other programmes, which are national in scope, the RIP has a regional dimension.

Support Schemes for Innovation Projects, the Introduction of New Products and R&D Projects

The schemes which are fostering the introduction of new products, R&D and other immaterial investments are as follows: the Italian law 46/1982; some schemes falling under the regional laws in Lombardy (e.g. L.R. 34/1985, L.R. 35/1996); the SMART programme in the United Kingdom; the FFF, ERP and ITF in Austria; some measures of the NT programme in Norway; the 'interest-free revolving loans' in Wallonia); and finally, some measures of KIM in Limburg.

Regarding the introduction of new products, the Italian Law L. 46/1982 is probably one of the most well-known incentive schemes that promote the transfer of know-how from research institutes to SMEs. However, because of administrative difficulties linked to the application process, this scheme has mainly benefited larger firms.

In Lombardy, the L.R. 34/1985 and L.R. 35/1996 schemes are also concerned with the development of new products and services. The Regional Law L.R. 34/1985 aims to stimulate increased productivity and improvements in the competitiveness of SMEs through fostering access of SMEs to research institutes, university departments and laboratories, as well as business services. Support is concentrated on applied research and specific innovation projects initiated by SMEs. Companies could obtain a subsidy covering up to 50 per cent of eligible costs with a maximum of €200000. Financial support is provided by means of grants, which were allocated through the CESTEC (SME Technological Development Centre) to companies collaborating with specialized institutes to develop, test and implement the results of common research projects. The scheme also foresees subsidized loans for innovative projects developed by individual firms through a special fund ('Rotation Fund for Innovation'), established by the Lombardy Region and collaborating banks. Because of complicated administrative procedures, the scheme remained relatively unsuccessful. However, in 1993, a new Regional Law was introduced (L.R. 7/1993) with the aim of facilitating procedures. Grants were provided instead of subsidized loans for research projects carried out by small handicraft enterprises. In addition, 'participating loans' to SMEs were introduced, whereby interest rates charged to the firms were linked to the profitability of the projects carried out – a novelty in Italy.

L.R. 35/1996 is based on experience accumulated from other schemes during the last decade, and involves a complex programme and mix of instruments and actions addressed to different sectors of the economy (industry, tourism, trading, and services). The scheme, launched in 1996, aims at (a) the promotion of service centres, technological poles, intermediate structures of support for technology transfer to SMEs; (b) direct support (through grants) to new firms working in engineering and production of new products and services (for research equipment, technician training, engineering personnel, technical consultancy and prototype development and testing); (c) grants for carrying out R&D projects, for participation in EU research programmes and for internships of graduates in R&D projects; and finally, (d) subsidized loans for innovation, including R&D equipment purchases. Support for innovative products and services is provided mainly by means of grants, access to the 'Rotation Fund for Innovation' and through medium-term grants or 'participation loans'.

In other European regions, there are a range of financial schemes operating through grants, loans and subsidized interest rates. In Denmark, the involvement of the State in the provision of risk capital has been based on the so-called Development Corporations and the Growth Fund, which provides loans for R&D. The two organizations are in the process of merging. The number of applications and projects approved by the Growth Fund has been decreasing steadily in recent years. There are 19 Development Corporations at work in Denmark with a regional or sectoral profile. Compared to the Growth Fund, the Development Corporation has a closer involvement in the day-to-day management of the subsidized firms. An important weakness of the Development Corporations is that they have no links with regional development agencies, and seem not to be well integrated into the area.

In Austria, there are three different innovation and technology funds. The Austrian Industrial Research Promotion Fund (FFF) supports research and development projects and concentrates on the early phases of the innovation process. It is interesting to note that the scheme follows a strategy of co-operation in international research through Austrian subsidiary companies, and that the programme is trying to help companies to evolve from component suppliers to system suppliers. The FFF pursues a bottom-up strategy, and firms decide for themselves which technologies and markets they want to explore. Through grants, loans and low interest rates, the fund concentrates on high-risk projects. The scheme stimulates R&D co-operation with universities, research organizations and technical colleges. However, most of the funds are still allocated within standard programmes, which are dominated by the linear approach to innovation.

The Technology Programmes of the ERP Fund relevant to this analysis comprise the Technology Programme and the SME Technology Programme. The programmes focus on the transfer of prototypes into regular production through the provision of loans at low interest rates for an amount of up to 50 per cent of total investment. Special banks (Treuhand-Banks) select projects and prepare the proposals, which are usually accepted within a period of two months. In general, ERP support aims at 'lower-tech' industries, and firms belonging to higher technological levels belong more to the clientele of the FFF. The ERP pursues a bottom-up strategy. The programme follows legal instructions, and there is not much space for innovative initiatives.

The Innovation and Technology Fund (ITF) only grants support for certain technological areas. In contrast to the FFF and the ERP fund, it has a top-down approach, trying to define those areas with the highest potential for technological development and economic growth. Innovative projects are supported through the provision of grants. There is a strong concentration of funds on larger companies, and industries that most frequently benefit from ITF grants show a 'high-tech' bias, mainly in the area of data processing, information and communication technologies. The major problem of the programme is that it is mission-orientated and it has too little funds available to grant significant support for each of its focus areas. Although the programme has been reorganized in 1999, it still follows the same top-down support strategy and similar goals.

In Wallonia, the Interest Free Revolving Loan aims to fund industrial applied research and development. It covers 70 per cent of research costs for SMEs and 50 per cent for large firms. It appears that the loan scheme is primarily targeted towards 'research rather than development'. There is too much focus on the promotion of high-technology products (with an uncertain market potential) compared to innovation projects, which may turn out to be 'profitable economic activities in Wallonia'. In fact, high-tech firms with a good regional image seem to enjoy an easier access to funds than firms in traditional sectors. Moreover, during the phase that follows the actual R&D, the number of support schemes available is rather limited, which makes the 'development phase' of the products all the more difficult. Another problem is the time lag between the application and the allocation of the subsidy. For many firms in sectors with short product life cycles, the subsidy comes too late. For other companies (the strongly innovative ones), the subsidy comes as an additional source of finance with little or no behavioural impact. Overall, a more integrated design of innovation support is necessary, allowing easier links between instruments or access in the form of a package.

5.2.2 Support Schemes for the Introduction of Technical Personnel in Small Firms (A2)

An important barrier to the introduction of innovation in SMEs is the lack of technical competencies. Some policy schemes address this problem through specific measures fostering the introduction of technical personnel in small firms.

The most interesting example of measures which foster the introduction of higher technical and professional personnel into SMEs can be found in Wallonia. The Technology and Innovation Manager (RIT) programme is exclusively targeted at SMEs. The aim is to strengthen resources for the development of innovation projects. The scheme covers the funding of a technological audit of a project and 80 per cent of the salary of an innovation manager, for one year. This programme is particularly relevant for firms which are not particularly innovative. One difficulty of the programme is the recruitment of a good innovation manager for the limited period of one year. The scheme, which was popular in the 1980s, has attracted little interest in recent years. This is due to the existence of other alternative instruments in support of human resources. However, the scheme has been relatively successful in the sense that it benefited non-innovative firms, which have, in most cases, been keen to extend the services of the RIT. For already innovative companies, the programme was considered more as a training and employment support programme. The value added by the scheme is therefore much higher in low-tech and non-innovative companies (SMEs). Overall, the RIT has a potentially high behavioural impact on SMEs. The employment of an innovation and technology manager often marks the start of a continuous innovation activity, fostering additionality in the innovation process.

5.3   SYSTEM-ORIENTED INNOVATION SUPPORT SCHEMES (B)

This section deals with innovation schemes which are oriented towards the productive system as a whole. We review the main cases identified in the studied regions.

B1   Policies Based on Technology Centres and Schemes Fostering Technological Diffusion to SMEs

These schemes are the most common schemes in support of indirect technological diffusion for SMEs. They can be categorized either as 'top-down' schemes whereby technology transfer initiatives are supply driven as in the case of Tecnopolis in Apulia, or as 'bottom-up' actions, whereby a group or network of small local firms tries to find solutions to existing technological challenges (the experiences in the Valencia region and to a certain extent in Lombardy illustrate this point). There are several examples of technology centres of both types, some being more effective than others.

In Norway, the RUSH (Regional development programme between state-owned colleges and SMEs) and the REGINN (regional innovation programme) are regionally-based schemes aimed at improving collaboration between colleges and local firms. RUSH was an experimental programme which lasted for four years and covered four colleges. This open-ended pilot programme became the REGINN programme in 1997. The programme has a double aim: assisting SMEs in need of external R&D resources and helping R&D institutions and colleges to strengthen their competence on SMEs' innovation needs. RUSH was regionally-based, focused on manufacturing SMEs, and allowed funding of 50 per cent of joint development projects to be funded, while provisions were also made for training and education activities. The original idea behind the programme was the Steinbeis system of technology transfer in Baden-Württemberg. The REGINN programme has a much wider scope, aimed at promoting and developing a regional innovation system, with particular emphasis on selected clusters of firms in the region.

The IMPIVA (Instituto de la Mediana y Pequeña Industria en Valencia) was created in 1984. It is a public organization in charge of designing and implementation of industrial policy for SMEs in the region of Valencia. The organization operates as a network which provides services to firms. The 15 Technological Institutes affiliated to IMPIVA offer information and documentation, technical studies, laboratory tests, consultancy, technology transfer and human resource training. The Business Innovation Centres promote new economic activities and stimulate the creation of innovative projects; the

Technological Parks promote investments in high technology industries and links between research institutions and firms. Four technological institutes are studied in SMEPOL: the Technology Centre on Ceramics (AICE) focuses its activity on R&D projects and technological consulting or transfer in the ceramic sector; the Technology Centre on Textiles (AITEX) focuses on human resource training and the production of multimedia training products for the textile sector; the Technology Centre on Toys (AIJU) is specialized in training in the toys sector; and INESCOP provides information services to the footwear industry. Several factors have probably contributed to the relative success of the Technological Institutes: their governing body is composed of firms' representatives; their operations are increasingly self-funded (an average rate of 60 per cent for the sample under study); and finally, the centres are integrated into the social and economic fabric and have frequent contacts with firms, while at the same time being well connected to other similar international centres. One of the most important merits of the institutes is to have prepared companies to achieve improved quality and production standards, which were of paramount importance to sustain competitiveness of companies at a time when Spain was joining the European Community. In doing this, the institutes concentrated their activities on technological advice, paying relatively less attention to joining R&D projects with local companies. Overall, the institutes have allowed firms of the region to evolve from a situation of 'imitators' to 'adapting SMEs' with some innovative capacity.

In Lombardy, L.R. 33/1981 provided support for the promotion and building of specific structures and facilities for technological assistance to SMEs, such as Centro Tessile in Como and AQM in Brescia.

In Apulia, Tecnopolis in Bari is a typical 'top-down' structure of a large (over 200 employees) technological park based on a supply-side policy aimed at offering existing academic expertise. Created in 1984, it was the first technological park created in Italy. The activities of the centre concentrate on technological transfer, applied research in industrial automation, training for innovation, supply of innovative services to firms and the public administration, participation in international programmes in applied research and the creation of new firms. One of the weaknesses of the centre is that is has never really engaged in a thorough analysis of the potential needs of local or external firms in order to make the interaction between research centre and economic activities as meaningful as possible. In fact, the centre is a typical 'self-refereed' institution, which finds it difficult to have its own performance evaluated by external examiners. Moreover, the way the centre is financed may also be to blame for the situation: 62 per cent of the budget is financed by national bodies, 29 per cent by regional and local institutions, 2 per cent by the EU and only 7 per cent by the private sector.

Technology centres and related institutions in Upper Austria have been conveniently categorized in three groups: R&D-orientated centres, facility-orientated centres and technology transfer centres.

In the first category, the Software Park Hagenberg offers an interesting example of a thriving institution linking together research laboratories, firms and technical colleges. The park also fulfils the role of an incubation centre, since many new firms are spin-offs of former research projects. Two major Austrian electronic companies are also hosted on the premises. The park is funded by the federal university budget as well as by contract research. The intensive interaction between researchers, engineers, managers and PhD students is considered to be one of the major advantages of the park. Interestingly, companies tend to collaborate rather than compete in order to succeed in their application for research grants.

A second important centre is the Research and Training Centre for Labour and Technology (FAZAT) located in the old industrial area of Steyr, which has been going through a process of restructuring in recent years. Originally designed as an incubation centre, it has become a technology centre coupled to a technical college where the activities concentrate on automation and telematics. Technology transfer takes place in larger companies located in the region, while relations with SMEs are less developed. The centre was set up with different kinds of public subsidies and it currently receives 50 per cent of its funds from contract research and consulting services. The project is considered as a successful venture, partly because of the political consensus which has supported the municipality in its efforts to set up the centre.

In the category of facility providers, we identify four different technology centres. The Incubation and Technology Centre, Wels (GTZ) provides infrastructure for new small firms and information, marketing and PR services, including help for R&D grants applications. The centre is more or less self-supporting and it is primarily controlled by private investors. The centre's performance as an incubator has been very satisfactory; the number of new firms created has exceeded all expectations. However, from the point of view of transfer of technology, performance has been very poor. The three other centres analysed in the study (Technology Centre, Linz, Technology Centre Innviertel and Technology Centre, Salzkammergut) are all focused on the provision of infrastructure and basic facility services. The fact that the largest centres in Upper Austria are not performing technology transfer functions constitutes a serious deficit in the technology and innovation system of the region.

Regarding the last category of technology centres, there is no central technology transfer agency in Upper Austria. However, the Innovation Relay Centre Austria (CATT) located in Linz focuses on certain aspects of technology transfer. The centre, which is a partner of the Innovation Relay Centre Austria and also a member of the European Innovation Relay Centre, offers active and passive technology transfer services. However, it is more an information provider than a consultant; its perspective is the European innovation exchange network, rather than the regional market.

In the UK, the London Lee Valley Centres (LVCs) and the London Lee Valley Business Innovation Centre (LVBIC) at Middlesex University were established by local agencies and authorities and supported by European and national funds.

The LVCs were established in 1995 with support of European Regional Development Funds (ERDF). They include four semi-independent centres in the field of design, telematics, technology transfer and teaching. The four centres aim to help businesses to expand and become more competitive and profitable and to stimulate industrial and commercial growth in the Lee Valley region. One of the major problems of the centres is that they are 'funding-driven', which means that the needs of the firms have not been considered, and projects are created to absorb available funds. This problem has been exacerbated by the requirements of quantitative output indicators prescribed by the ERDF. Consequently, the projects lack a clear conception of the potential demand for services, the available university facilities and staff and a perspective on how the centres relate to other business support initiatives within the region.

The LVBIC is one of the 140 centres sponsored by the European Commission DGXVI in less-favoured regions of the Union. The centre focuses on the provision of intensive support for selected innovative individuals and enterprises, with emphasis on technological opportunities and product innovation, including the commercialization of innovative ideas. A key component of the support package is the provision of finance in the form of a short-term loan or as equity. The centre also supports firms applying for government schemes (such as SMART for instance) and it has its own annual award scheme, the 'London Lee Valley Innovation Award'. The most important weakness identified by clients is the insufficient financial support and the lack of access to venture capital for innovation.

B2   Schemes Fostering the Role of 'Innovation Brokers'

These schemes distinguish themselves from other schemes in this section because of the active role played by 'technology or innovation brokers' who identify technological needs of firms, even when these are not yet explicitly expressed by entrepreneurs. Examples in the studied regions include the Syntens scheme in Limburg, the Business Links programme in the UK, the

TIC and GTS schemes in Denmark, and also some measures of the Norwegian TEFT programme.

One of the best examples within this typology is Syntens in Limburg (the Netherlands). Syntens stems from a previous experience when the Dutch government established a network of 18 Innovation Centres back in 1987. Each Innovation Centre had a consulting staff of 5 to 10 people, mainly engineers who actively contacted and visited SMEs in the region. The role of the ICs gradually changed from bringing technology to regional SMEs to an intermediate role of 'broker', and more recently fulfilling the role of 'organiser, animator or coach'. The IC network privatized in 1993 and merged with the IMK (Institute for SMEs) in 1998, and was renamed 'Syntens: an innovation network for entrepreneurs'. The main task of Syntens is to raise the awareness of continuous innovation and input of new knowledge amongst firms in the region.

In the UK, the Business Link scheme provides assistance for innovating SMEs, following the concept of a 'one-stop-shop', encouraging firms to access external support through a single channel. These services include personal business advice, diagnostics and consultancy services, marketing training and advice, international trade advice, and financial information. A Personal Business Adviser is assisting SME entrepreneurs in defining their business needs and formulating their strategy. These generalists are assisted by specialist counsellors including Innovation and Technology Counsellors (ITC) and Design Counsellor (DeC). Business Link adopts a holistic view of business development, providing an integrated range of services to clients, in which technical advice is rarely provided on its own, helping firms to reap the full commercial benefit of their innovations. Often, client firms can establish an ongoing relationship with their assigned Business Link advisers. While smaller firms are often recipients of generic assistance, larger and older firms are likely to be referred to external consultants. The networking component of the scheme is rather important, since Business Link is concerned with linking SMEs to elements of the external support system – mainly consultants – addressing above all the commercial dimension of their innovations, rather than issues related to product or process development.

The Technological Information Centres (TIC) in Denmark constitute a network of advisory centres, which are based in 15 different regions of the country. They deliver information services to SMEs in search of specialized technical advice. The TICs operate as broker between technology-orientated SMEs and a variety of management and technical consultants as well as research institutions relevant to SMEs. Services include the dissemination of information, coaching and regional co-ordination of advisory services. It is interesting to note that all 15 centres are managed by a national centre called TIC Denmark, which ensures the existence of an overall national strategic framework of action to co-ordinate the initiatives of the 15 regional centres. In addition, TIC Denmark is also involved in policy negotiation with the national government and, thus, it is in a position to balance national and regional interests within the overall support system. However, this has been subjected to problems in recent years, the TICs finding it difficult to respond to the specific needs of smaller business communities and at the same time expand the scope of target industries and service tasks within given budgetary limitations (together, the government and the counties contribute almost 100 per cent of the budget). The situation seems to be evolving towards more regional autonomy for each centre.

The Danish Approved Technical Services (GTS) comprise a large number of formerly independent research institutions and industrial service providers which still operate as independent business units. The GTS Institutes operate as a link between the research community and the business community; they are involved in the development of new technologies and the diffusion of existing know-how originating from other sources. It is interesting to note that the GTS Institutes are private businesses with goals of their own, but at the same time they provide a public service for which they are paid by the State. Inevitably, this creates dilemmas in the operation of these institutes. Technology services provided by the centre are seen in a linear perspective; the whole network could benefit from a better understanding of clusters of production, value chains, networks of personal relations, tacit knowledge and the functioning of a regional innovation system at large. While the TICs are caught in a dilemma of regional versus national programme co-ordination, the GTS face their own dilemma between the needs of market-based large clients and the specific (less economical) needs of small firms.

B3 Mobility Schemes for Researchers

The mobility schemes promote the diffusion of technological expertise and the ability to manage newly acquired know-how, through the mobility of researchers between different organizations and institutions such as technology institutes, universities and private firms. The indirect aim of such programmes is to cluster relationships between the above actors and to promote the transformation of researchers into economic actors and entrepreneurs. In other words, it is hoped that a researcher or a scientist who has the chance of working for a company on a given project, will eventually adopt an entrepreneurial perspective, or will at least establish new links between the firm and the home institute.

In Wallonia, the scheme FIRST enterprise (Formation et Impulsion à la Recherche Scientifique et Techologique) aims to strengthen the scientific and technological potential of firms through the employment of young researchers working part-time in the company and part-time in a research laboratory. The programme emphasizes the aspect of 'knowledge creation' rather than the industrial application of this knowledge. However, the original feature of the programme is to promote collaboration between research and high-tech companies. The number of applications has always been relatively high, and the recruitment of FIRST researchers does not seem to be a problem. Through the scheme, companies have access to external equipment, material and competencies, and these resources are often a prime motivation for accessing the FIRST programme. Also, the scheme may help establishing a routine in linking with external resources of technologies and expertise. This includes informal contacts, which are very important for the company's 'technological watch'. Overall, the programme offers the companies the possibility of acquiring qualified researchers; it provides privileged access to specialized equipment, and establishes an 'industry–university interaction' which may yield significant benefits in the long term.

The KIM programme in Limburg (which stands for Knowledge-carriers in medium and small-sized firms) focuses on the mobility of human resources. The scheme aims at stimulating and supporting small firms to engage in technological innovation through special funding, through the reduction of the cost of hiring a highly educated graduate for one year. The idea is to transfer technology and to diffuse technical and management skills from institutions of higher education to SMEs by using young graduates as knowledge carriers. The major outcome of the KIM is that it has lowered the barrier of employing personnel with higher education. KIM is based on an interactive mode of intervention, whereby a consultant acts as a mentor or coach to the firm, translating informal interaction with the firm into a formal plan of action. The consultant 'actually decodifies the policy instrument and uses its codifying skills to codify the tacit needs of the firm' (Nauwelaers et al. 1999). The scheme, which addresses the internal innovative capabilities of firms, is perfectly suited to firms, which operate within a logic of 'learning-through-producing'. Overall, participating firms seem to share a more positive attitude towards highly educated employees and innovation. Also, there seem to be other systemic effects internal to the firm, but external to the supported innovation plan.

In Norway, the NT programme has introduced a scheme called 'Technology mentorships' which allows the involvement of scientific researchers in solving specific technological problems arising from innovation projects of SMEs.

In Lombardy, L.R. 35/1996 offers grants for internships for newly graduated students in the field of R&D activities. This particular measure could have been seen as a coherent tool to improve the social capability of firms to innovate through the placement of technical staff in SMEs and the building of privileged links between firms and universities and research centres. The reality turned out to be quite different. Indeed, the number of accepted applications has been extremely low so far, demonstrating that being in keeping with a theoretical line of action is no guarantee of success.

5.4   PROCESS-ORIENTED SUPPORT SCHEMES (C)

If one takes into account the main outcome of the literature on industrial districts (Becattini 1987, 1998; Brusco 1982, 1989; Garofoli 1983, 1989), models of local development (Garofoli 1991; Storper and Harrison 1991), endogenous development (Garofoli 1992), learning regions (Morgan, 1997; Maskell et al. 1998) and regional innovation systems (Gaffard 1992; Asheim and Cooke 1999), it appears clearly that the only way to foster the competitiveness of local systems of small firms, especially in regions and countries with higher living standards, is a strategic transformation based on quality and innovation. This 'high road to development' (Pyke and Sengenberger 1992) is based on the fundamental pillar of strengthening external economies (external to the firms but internal to the territory) (Garofoli 1983, 1989).

However, it seems that, as yet, relatively few regional policy-makers have taken on board lessons emerging from theoretical reflections and interpretative analyses of regional and technological change. The support schemes analysed above are fairly conventional: they are mainly static and operate in a one-way direction; there is little additionality; and there are few interactive or dynamic relationships.

There are a few examples among the study regions of interactive tools which can induce the accumulation of localized learning and the strengthening of the fabric of relationships among local actors and institutions (along the lines of the main arguments in Chapter 2).

These instruments can be categorized in two groups. The first group (C1) concerns interactive tools based on proactive actions and initiatives. These tools are most appropriate for the promotion of continuous interactive learning through the implementation of networking structures at the local and regional level. They are process-oriented. The SMEPOL project identified two interesting examples of such actions in Norway. The second group (C2) covers schemes aimed at upgrading the local and regional innovation systems. Two schemes, one in the Netherlands and the other one in the UK, fall into this category.

C1   Interactive Tools Based on Proactive Actions and Initiatives

The TEFT (Technology Diffusion from Research Institutes to SMEs) programme aims to help SMEs in manufacturing and services to collaborate with the four largest polytechnic research institutions in Norway. The idea is to draw non-R&D-intensive SMEs into the national innovation system and to make SMEs become continuous customers of the national R&D system. The scheme is further enhanced by the action of county-based technology attachés whose task it is to match technological needs of firms with the technological potential of the institutions. In this respect, the TEFT programme is essentially a proactive scheme in which technology brokers aim to lower existing barriers of co-operation between institutions and SMEs through frequent visits. The scheme comprises the identification and subsequent subsidization of technology projects for an amount of 75 per cent of the total cost, the contribution of TEFT being used to buy services from selected institutions. One of the weaknesses of the programme is that, to some extent, it is supply-orientated, serving the interest of participating institutions first, by focusing on the generation of income through projects. In fact, it is sometimes unclear whether the programme's priority is to improve attitudes of research institutions vis-à-vis firms, or whether it aims to increase SMEs' use of R&D competence. Finally, the issue of achieving innovation and competitiveness without R&D should also be addressed. Indeed, it is questionable whether low R&D-intensive SMEs need the kind of policy instruments provided by the TEFT. Innovation, as already extensively developed in previous chapters, is not only about R&D.

The NT-programme (Innovation and Technology programme for Northern Norway) started in 1987 and is a regionally-based programme. The aim is to provide financial support to projects in firms of the region, to strengthen the co-operation between centres of expertise and the firms, and to strengthen the co-operation between firms and within firms. The main focus of the scheme is to provide funds for innovation projects, but the philosophy is to provide all-round proactive support for innovation (that is why programme managers are following firms closely and establish long-term links with them). The approach is tailor-made and intends to meet the specific needs of firms (there is a system of technology advisory contracts with R&D institutions). The target group of NT are R&D-intensive firms or the 'best' firms in manufacturing and consulting. The programme foresees support for the development of products, production processes, marketing and collaborative links between firms and R&D institutions. It has no 'infrastructure aim' but concentrates more on the relations between firms and institutions. From a general point of view, it seems that NT has been performing a co-ordinating role in the support system in Northern Norway.

C2   Schemes Aimed at Upgrading the Local and Regional Innovation Systems

The second group concerns schemes aimed at upgrading the local and regional innovation systems. SMEPOL case studies of such programmes include the KIC – Knowledge intensive Industrial Clustering – in the Netherlands, the new innovation policies in Italy, and the SMART Network Clubs in the UK. Other well-known examples include the cases of Grenoble and Toulouse, where public and private actors intervene together to improve the innovation capability of the local system, through programmes which enhance interaction and accumulation of local knowledge.

The literature on local development emphasizes that the best way to support the start-up of an innovative local process is through the upgrading of the technological knowledge within the local system. The policy schemes which try to introduce better knowledge resources are, then, goal-oriented.

In this category we find all schemes which are concerned with the systematic upgrade of the social capability of a local system to foster and control the development of technology. The idea here is to facilitate the diffusion, within the local and regional system, of a collective ability to promote change and innovation within the firms. In concrete terms, this refers to any actions that may lead firms to improve their attention to quality when acting as suppliers. Examples include the improvement of services provided by technological institutes, the development of new links between university and industry, the systematic upgrading of human resources and managerial skills in support of innovation, and finally the improvement of the awareness of local and regional actors regarding training and development of new professional skills.

The main distinction between the present category and previously-mentioned categories relates to the final recipient of the measure. The present scheme is directly aimed at upgrading the innovation capacity of the local productive system as such, whereas the previously-mentioned schemes may (or may not) achieve this same goal, after the creation and implementation of interactive tools. In essence, these schemes are concerned with the upgrade (in a systemic fashion) of the social capability of a local system to foster and control the development of technology. The idea here is to facilitate the diffusion, within the local and regional system, of a collective ability to promote change and innovation.

The KIC (Knowledge-Intensive Industrial Clustering) is a joint public–private initiative aimed at upgrading regional SME suppliers through co-operation with each other in co-engineering projects for the renowned Dutch multinational OCÉ. Most KIC projects were initially set up to engineer systems or modules for OCÉ's new colour copier. The multinational company was interested in outsourcing manufacturing and engineering to a group of co-operating suppliers within the region, to turn regional 'jobbers' to 'co-makers' with the help of the Dutch innovation centres 'Syntens'. The scheme has been particularly relevant in Limburg where many industrial SMEs were 'isolated' and unable to use their skills in a codified way. Moreover, the use of private R&D facilities helped to compensate for the relatively weak public research infrastructure. However, one of the weaknesses of the scheme was that the initial 'shared collective goal' from upgrading regional SMEs co-makership evolved towards upgrading OCÉ's suppliers into main supplier. Having said this, firms have learned to codify their tacit skills and to communicate knowledge with OCÉ's R&D unit, to co-operate in engineering with other SME suppliers in the region, which in the end favoured the emergence of micro-clusters which acquired dynamic competencies that outperform the competencies of individual firms. On the negative side, we note the existence of complex procedures and an insufficient training content of the programme. The project also suffers from being 'locked' into the technological trajectory of OCÉ and, therefore, there is a need for other leading firms to reduce the domination of the multinational company.

The industrial and innovation policies introduced in Italy through the national law 317/1991 constitute, to a certain extent, a novelty. The new provision makes it possible to draft industrial policies which are in direct support of the territory and the local productive system as a whole, stressing the interdependencies that exist between decisions of firms located in industrial districts. Thus, new innovation schemes have been introduced in industrial districts to support collective initiatives and consortia of small firms (cf. the schemes introduced in Lombardy, through the regional laws 7/1993 and 35/1996), strengthening thereby the local system as a whole.

Support schemes for start-ups are also aiming at the upgrading of local and regional innovation systems. These are all the instruments which foster the creation of new firms on the basis of acquired technical know-how in industry or academia.

A good example of a start-up scheme is the Small Firms Merit Award for Science and Technology (SMART) which was introduced in the late 1980s by the UK Department of Trade and Industry. The scheme is based on an annual competition, and targets innovation support through feasibility studies and development projects for potential entrepreneurs at the pre-start-up stage, or very young micro-enterprises which are highly R&D-orientated. Because the scheme is only targeting a certain class of firm, it is not always able to contribute to raising the overall level of innovation of SMEs in the region. While the programme may have had a positive impact on the ability of entrepreneurs to proceed with innovative projects, commercial objectives have proved more difficult to achieve. This weakness calls for a more holistic approach and for the provision of marketing assistance and other business support measures to successful firms. Finally, in view of a longer-term strategy of support, the recent creation of regionally-based networks of SMART winning firms may become privileged fora for interactive learning.

5.5   POLICY CONCLUSIONS

A major outcome of the SMEPOL analysis is the necessity to further develop the awareness of SMEs for continuous innovation and change (see Chapter 4). Programmes serving this goal need to be implemented in an interactive fashion, using all positive externalities that the local and regional productive system can offer. This includes networking between the regional production sphere and research institutes, as well as closer collaboration between public and private institutions alike. Above all, such programmes of continuous education will ensure that sufficient pressure is put on the firms to engage in a collective fashion in a constant effort to renew their knowledge base. This idea has been implemented in Wales, where the Welsh Development Agency, through its Employment and Training Centres, has set up a programme which promotes the development of a learning organization at the regional level (Cooke and Morgan 1998).

   Overall, there is a need for an interactive process generated by a 'bottom-up' approach for the promotion of competitive advantages based on localized knowledge and learning. This calls for enhanced relations and networking: networking at the level of firms (clients, subcontractors and suppliers), institutional networking (research institutions, services and technological centres, training institutes, chambers of commerce), and networking between public and private organizations (associations of producers and other stakeholders). Together, these partners should produce new expertise and competencies which are external to the firms but internal to the local system of firms. In other words, there must be a strategic innovation planning to foster the internalization of external codified knowledge and the adaptation, development and diffusion of existing tacit knowledge, through the wealth of relational assets of the firms. Through the implementation of interactive tools, this strategic planning must, at the same time, use all instruments which can help bypass the technological barriers faced by SMEs.

To achieve this, the collaboration between private and public actors must facilitate the development of a project culture, which also includes the ability to evaluate and correct existing schemes, from a systemic point of view. In the pursuit of such collective objectives, destructive competition to maximize individual gains from public financial resources should be avoided when possible. Instead, the new schemes must emphasize the creation of interactive processes, which may deliver additional results and additional investments for every individual partner, and for the system as a whole.

In concrete terms, this means that policies will have to concentrate on 'client-focused services' where intermediate structures have the ability of transforming the tacit needs of SMEs into explicit demand; on actions that will introduce new expertise into SMEs; on programmes that will promote interface structures between industry and research; and, finally, on schemes that will gradually involve research and higher education into SME problem-solving exercises. The next chapter discusses how far the SMEPOL instruments are able to meet these goals.

To conclude, what is at stake here is a systemic approach to co-ordinate innovation and industrial policies human resources and training policies with the final aim of introducing some form of strategic local planning.

6. Results and impacts of policy instruments

**Javier Alfonso Gil, Antonia Sáez-Cala,
Antonio Vázquez-Barquero and
Ana Isabel Viñas-Apaolaza**

6.1 INTRODUCTION

The objective of this chapter is to determine the effects and impacts of the various SMEPOL government-implemented policies with the aim of improving innovation in SMEs. Evaluating policies and describing results and impacts is common to any action in economic policy, i.e., we are dealing with means and ends and how the former influence the latter. If measurement shows that the objectives are attained, it can be asserted that the tools used were coherent and positive.

One of the greatest difficulties is that firms, far from being static, undergo constant change. If the tool used tends to remain relatively static due to institutional design, the possibilities of achieving its objectives decrease because of a growing gap between the relatively static institution and the evolutionary path of the firm. Therefore, the policies and instruments created and designed must be institutionally flexible in order to adapt to the firm in its evolution over time, as well as adjusting to different types of firms, in terms of the size of the business and the sector that it is in. The next chapter discusses this in more detail.

Firms are aware that to exist in the market requires continuous improvement of processes and products and that the best way to achieve this is through their own effort and resources. The SMEPOL reports confirm that the firms analysed, active agents in competitive frameworks, generally reach their objectives of innovation and adaptation to technological change by means of their own resources (Chapter 4). Therefore, innovation support made available by the public administration is important but secondary. However, the fact that these policies are secondary to firm innovation does not mean that they are not relevant or decisive. On the contrary, in many cases, they may actually determine the success or failure of technological change in firms.

The sources of information of this chapter are the seven national reports of the SMEPOL project. The various policy instruments have been assessed using surveys (telephone, postal and interviews) of sample groups of clients, in-depth interviews with selected clients in some cases, and interviews with administrative and business support staff involved with policy implementation. Some studies have also made use of 'control group' surveys of firms, particularly where some of the firms included have been beneficiaries of the instruments. In most cases it has also been possible to draw on pre-existing studies conducted by government evaluation units or by consultants/academics.

There are a number of key methodological issues affecting the appraisal of policy instruments. The first relates to the difficulties involved in attempting to quantify the actual impacts of policy interventions on the firm. There may be a significant time lag before the full effects of an intervention become apparent. With respect to technology-based innovations in particular there can a considerable period of time (invariably several years) between the inception of the supported project and the commercialization of the product. Furthermore, it may be difficult to isolate the real effects of the policy intervention (for example the additionality[4]) from a host of other factors affecting innovative performance. In some cases firms may have been beneficiaries of a number of policy tools, further complicating evaluation. Hence, quantitative data on policy impacts need to be supplemented by the qualitative judgements of client owner/managers themselves. Qualitative indicators are also important in terms of capturing the extent to which policy instruments are contributing to the capacity of businesses to undertake innovation. In this respect an important process indicator relates to the frequency and type of interchange between agencies and businesses, particularly with regard to the extent to which mutual 'learning by interacting' is being fostered. Finally, in the absence of detailed data on longer-term impacts on client firms, the design of policy instruments and their manner of implementation can be examined with reference to models/elements of good practice identified by previous studies on innovation and SMEs.

This chapter considers the results and the effects of the policy tools on their primary objectives, measured by: level of awareness, use and level of satisfaction of the policy tools; target firms by size, sector and technological level; adjustment between policy tools' supply and SMEs' needs; and policy instruments' contribution to innovation activities and performance within SMEs' clients. The impacts deal with the effects (considered as longer effects) that the policy instruments have induced outside the client firms, encouraging linkages and interactive learning, both between firms and between firms and the innovation support/knowledge infrastructure.

6.2 EVALUATION OF INSTRUMENT OBJECTIVES ACCORDING TO RESULTS

6.2.1 Results and Impacts for Direct Support Schemes

The majority of the SMEPOL policy tools are related to the financial barriers for SME innovation. In this section the following direct support instruments are included: FFF, ERP, ERP–SME, RIP and ITF from Austria; Interest-Free Revolving Loans in Wallonia; Growth Fund and Development Companies in Denmark; Italian National and Regional Laws; NT from Norway and SMART from the United Kingdom.[5]

Aims and target outputs of direct support schemes
The Austrian policy instruments (FFF, ERP, ERP–SME, RIP and ITF) are direct support programmes and target innovation, technology and R&D in firms. The FFF focuses on the beginning stages of the innovative process (applied R&D). However, its design includes more specific goals such as the promotion of international co-operation in the area of research, aid for SMEs and support for industries that, although strategic to Austria's economy, lack significant R&D programmes. The Technology Programmes of the ERP Fund selected are the Technology Programme ERP and the SME Technology Programme ERP–SME. Their objective is not research but rather adopting and adapting technology. They target the commercialization of SME innovation, particularly the transfer of prototypes into regular production. It is open to firms of any sector, industry or region of the country. The final goals of RIP are the reconversion of declining industrial areas, structural improvement and economic growth of peripheral regions. Further, the RIP is not only an innovation support instrument but one that also encourages investment and job creation. The ITF centres on applied R&D.

The Interest-Free Revolving Loans in Wallonia were designed to finance R&D applied to industry through grants to innovative or potentially innovative SMEs. In practice the instrument tends to fund research rather than development projects, with the implicit aim of creating innovation with a higher degree of novelty.

The primary aim of Growth Fund and Development Companies in Denmark is to promote growth and development in firms, by improving competence, responsible and risk-sharing capital to the benefit of employment and sustainable development of society.

Italian Laws have been introduced in order to provide financial support for the development of innovative projects. L.R.34/1985 was the first law produced at the regional level providing direct funding to enterprises for innovative projects. L.R.7/1993 was introduced to modify L.R.34/1985 to differentiate the type of intervention in favour of smaller firms (less than 50 employees). L.R.35/1996 is the latest instrument of Lombardia that supports local entrepreneurs in the development of innovative projects. This law provides many different instruments which support SMEs along the innovation process.

Norway's NT programme was designed to create new activities in firms in Northern Norway showing innovative skills. To achieve this goal, NT funds projects that strengthen co-operation among firms as well as between firms and knowledge institutions, thus improving firms' competitiveness.

The United Kingdom SMART support programme for innovation is centred on the development of new products, processes and prototypes and the stimulus of marketable innovative technologies in SMEs. It gives support to potential entrepreneurs in the initial stages of firm creation and aid to recently established micro-firms oriented towards R&D.

The majority of direct support tools have expanded during the nineties. In Austria a significant growth applies especially to the FFF with regard to the funds accepted for project support and the RIP grants. In the case of the ERP, while the overall loans have increased continuously in the 1990s, the funding volume of the technology programmes (the general programme and the specific SME programme) decreased after a period of a very strong growth until 1993/1994, nevertheless providing clearly more loans today than in 1990. In Wallonia Interest-Free Revolving Loans represent 200 dossiers over 1995–97 relative to a total of 399 for all support instruments in the region and quantitatively becomes the most important support instrument.

Nevertheless, some direct support instruments decreased significantly. In the case of the Austrian tool, ITF, the number of applications, the acceptance of projects, the amount of funds granted and the scale of projects have decreased since 1994 due to the fact that the ITF is to be replaced by a new support programme. Regarding the Danish Growth Fund the number of applications, the number of projects approved and the incoming contacts have decreased over the years. Use of these innovation support schemes is also low in the Development Companies.

In general, awareness of direct support is low outside the client samples. Nevertheless, in Italy almost 93 per cent of the sample firms in Lombardia and 70 per cent of the surveyed firms in Apulia knew about the existence of specific instruments available to SMEs developing innovative projects. In Lombardia almost 71 per cent of the total answering firms applied for at least one of these tools, while in Apulia the percentage is higher (85.2 per cent).

Clients tend to have a favourable attitude towards the instruments and the majority of managers are generally positive and satisfied with direct support. Thus the general level of satisfaction with the support instruments seems to be quite high, just as most SMEs who have received support are satisfied with the benefits.

Most of the direct-support instruments, with the exception of ERP–SME, which is only accessible for firms of up to 250 employees, make no distinction between firms of different size categories. Nevertheless the proportion of SMEs in these instruments is high: in the case of FFF, SME-sized firms account for nearly 75 per cent of the supported firms. Most beneficiaries of the ERP programmes are SMEs. Except for the SME technology programme of the ERP – which is accessible exclusively to SMEs – the Austrian programme that focuses most on SMEs is RIP. Interest-Free Revolving Loans are predominantly used by SMEs.

Regarding the SME size categories, there is more diversity among tools. For instance SMART does not deliberately target very small firms with less than 10 employees, even though the majority of clients are start-ups or micro businesses. In the UK direct support schemes target very small firms with less than 10 employees, direct support in Denmark SMEs concentrates on firms of 10 to 49 employees and Laws in Lombardia are aimed at firms with 51 to 100 employees.

Many direct-support tools focus on high-tech and innovative firms. In Austria FFF supports high-tech projects and innovations which are more than incremental; ITF only supports innovative projects of existing and successful companies and these projects must have a sufficiently high growth potential. In Norway the main target groups of the NT are technology-advanced firms, R&D-based and new technology-based firms. In the UK SMART winners are highly R&D-oriented in terms of the proportion of current income dedicated to R&D and staff time allocated to this activity.

Certain technological areas like electronics and microelectronics, information and communication technologies, medical and optical instruments, chemicals, pharmaceutical technologies, advanced and new materials and business services are predominant in receiving direct support. That leads to a situation in which some industries are under-represented. In Wallonia for Loans most low-tech industries such as food, wood, textile, leather/shoes, and paper/publishing/printing are under-represented, especially in the light of their importance to the economy. In Austria, metal products, textiles, furniture and food are under-represented in the structure of the FFF support. In Norway, the fish-processing industry, an important sector of the Northern Norwegian economy, is outside NT's main target group of firms. However, other instruments show a large share of firms in the lower-tech sectors. In Austria, RIP and ERP are more relevant for lower-tech industries where machinery and manufacturers of metal and plastic products are strongly represented.

In the UK sectors which were particularly present in SMART award-winners do tend to be concentrated in manufacturing sectors, including manufacture of machinery and equipment, materials production and instrumentation and medical technologies. In the case of Denmark, client firms are predominantly from the stainless-steel industry. In Italy, data relative to the projects funded show the highest percentages in industrial automation, measure and control instruments and the development of other products.

Effects on client firms of direct support schemes

As far as product innovations are concerned, in Austria the direct support schemes are actually successful in stimulating innovation. According to the objectives of the support schemes, nearly all supported firms were able to introduce some form of product innovation. But also innovations, which are new to the market, are more frequent in the group of supported SMEs than in the SME sector in general. The most ambitious goal, as far as innovation is concerned, is to establish the technological basis for new products, most frequent in the case of firms participating in direct-support programmes, especially the FFF and the ERP Technology Programme. In Denmark, out of 85 enterprises involved in product and/or process innovations over the last three years, 19 enterprises have used innovation support schemes in connection with product development. Approximately 37 per cent of the enterprises have been involved in product innovations and only 12 per cent of these enterprises have made use of support schemes. The contribution of direct support instruments to process innovations is also weak in the case of Denmark (18 per cent of the firms have been involved in process innovations over the last three years and only 3 per cent have made use of support schemes). This also occurs in Austria, except in the case of the RIP programme.

A key indicator of success for product innovations, developed for the NT evaluation, is the degree to which these products succeed in the marketplace. Sales share accounted for by new or improved products during the preceding three years is higher among NT firms than firms on the average. The results are positive and firms report that NT has contributed to commercialization of a significant number of product innovations. The main contribution is the extent to which NT has improved abilities and opportunities within the firms in researching markets, integrating the innovative activity with the firm's broader strategies, achieving efficient project management etc.

Similar effects arise in the case of Interest-Free Revolving Loans in Wallonia which generate important effects on innovation experience and behaviour in firms. By implementing an R&D project, firms learn about the innovation challenge and success. In Austria also more than half of the respondents claimed that the direct support initiated a continuous or intensified innovation process which extended beyond the supported project, particularly in the case of the ITF. Nevertheless, the other half of the firms did not continue to innovate, but rather still regard innovation as occasional.

In Austria most of the firms answered that the innovation project or activity could not have been finished without support. However, the efficiency of direct support programmes regarding the stimulation of innovation may be rather low, because almost half of the firms (47 per cent) considered the support not really necessary. Firms in Wallonia would have conducted the R&D project without the Loans, but for most of them the implementation of such a project would have taken them much longer and also the financial help enabled the company to begin the project earlier and dedicate more resources to it, thus increasing its chances of success.

In Denmark more than 75 per cent of the enterprises found that support had no role in the initiation of the product innovation activity; in the case of process innovations, the pattern is even more extreme (90 per cent). It is difficult to find any significant relationship between firms' use of support schemes and their innovative activity. Thus, there is no sign of any relationship between firms' investments in innovations and their use of Growth Fund and Development Companies.

In Italy a higher percentage of firms in Apulia than in Lombardia made investments which, without support, would not have been carried out: 47.7 per cent of the respondent firms in Apulia said that investments planned in the funded projects were conditional on the support obtained, while the question in Lombardia obtained only 9.4 per cent of positive answers. In Lombardia the large majority of firms (90.3 per cent) would have carried out the investments planned without the funding.

The NT programme has a high degree of additionality for the one third of firms which reported that the project would not have been possible without support. Other firms answered that the project would have been delayed or implemented on a smaller scale. SMART offers additionality in two respects. First, it contributes to the development of new businesses that would not have started without the award: in 8 out of the 11 new businesses in the surveyed sample the venture would not have been able to start at that time without the grant. Secondly, out of the 29 existing businesses surveyed which received a SMART award, 35 per cent of the projects would have been completely abandoned, 52 per cent would have been delayed and 7 per cent would have gone ahead on a smaller scale.

Effects on business performance consider the overall performance of client firms, including employment, productivity, management and workforce skills, profits, sales and market shares. The performance of SMEs supported by direct tools has improved in various fields.

In the case of the UK more than half the managers surveyed reported that the performance of the firm had improved as a result of the SMART project. Even though nearly half said performance had not improved, almost all of these firms said they expected performance to improve at a later date. There is some evidence that SMART does contribute to improved performance, although it takes considerable time before results begin to appear, as occurs in other policy tools evaluated. When asked in what respect performance had improved, over a third pointed to increased profits either initially or in combination with increasing sales; 28 per cent identified some benefits from the project in terms of jobs saved and one third of firms referred to new staff being taken on as a result of the SMART supported project.

In Austria the creation of new jobs is favoured by most direct-support instruments. In particular the ERP programmes and the RIP consider the establishment of new jobs an important criterion for the project evaluation. These tools generated new jobs between 1990 and 1997. According to the official statistics of the ERP for 1997, all programmes together have stimulated 3201 new jobs, its general technology programme 825 and the SME programme 202. This is on average 15 jobs per project in the first two cases and 8 per project in the case of the SME programme. The RIP has generated 676 new jobs in this year which is on average 7 per project. According to the results of our survey 72 per cent of the SMEs which have received direct support were able to create new jobs due to the support. Also in more than half of SMEs, the supported innovation projects or activities have led to rising labour productivity while capital productivity is less frequently increased by direct support.

SMEs using some direct support instruments identify positive benefits arising in terms of management and workforce skills. More than three-quarters of SMART winners identified some positive effect in terms of management skills, particularly business planning. Furthermore two-thirds of surveyed firms reported that skill levels had increased in the workforce, which they attributed to the SMART-supported project. Through the Interest-Free Revolving Loans human resources directly involved in the R&D project implemented have also gained experience.

In developing overall performance in SMEs, expansion of the market share is also emphasized. In Austria this applies to all types of instruments, but internationalization and the diversification into new markets is more frequent in the case of firms participating in direct support programmes. A positive effect on internationalization seems to be a specific advantage of the ERP–SME Technology Programme. As a contrast, the expansion within the same market is more typical for other ERP programmes and RIP. Also access to an Interest-Free Revolving Loan has the potential to generate significant effects on market position. Thus, for successful projects, Interest-Free Revolving Loans have a strong impact on time-to-market and, in turn, on firm performance and market position.

Long-term effects and impacts of direct support schemes
It is difficult to evaluate the longer-term impacts of direct support instruments for reasons given earlier, in particular the time lag between the inception of the supported project and the commercialization phase, and the difficulty in isolating the real effects of the policy instrument from a host of other factors. Nevertheless, one way of gaining an impression of the wider effects of policy instruments is to focus on the extent to which they appear to encourage closer linkages and interactive learning, both between firms and between firms and the innovation support/knowledge infrastructure.

In addition the effects of the public funds obtained must be considered, due to the need to understand whether the funded projects gave new perspectives to the firms or just represented a means of reducing the financial weight of the project itself. The latter occurred in Lombardia where most of the firms obtaining at least one supported project said that public funding did not create new perspectives for the firm's activity. On the contrary, in Austria and Wallonia direct support generated important effects on firm innovation experience and behaviour. Links between firms and direct-support tools help SMEs in their future innovation activities. Interest-Free Revolving Loans help firms learn about the challenge of innovation and success.

Stimulating co-operation among firms is rather frequent (indicated by approximately half of the supported SMEs) in Austria, especially in the ERP–SME Technology Programme and the RIP which seem to be more effective than the other programmes in initiating co-operation with other firms. As to types of partners, continuous co-operation with research institutions is less frequent than with other firms. Moreover Austrian direct-support tools are successful in inducing spillover effects on other firms in the form of improved products or services.

Another possible impact on firms outside the user group occurs through the mobility of personnel, as in Norway, where employees working on NT-supported projects left to work in other firms. Another impact arises from the establishment of new firms that are in some way linked to NT projects. Almost half of the supported firms were involved in some form of spin-off. In a few cases new production technology diffused to other firms gave rise to increased earnings and competitive advantages for several firms.

Regarding links between university and SMEs, SMART has facilitated some spin-outs from universities in London. It is interesting to note that a recent innovation in the case of the SMART scheme is the introduction of network clubs at a regional level in an attempt to facilitate a sharing of experience between firms and to encourage co-operation.

NT is to an increasing extent matching funds with other programmes aimed at commercializing results from R&D and fostering a strategic basic research programme for marine biotechnology. NT staff provides such expertise from outside the programme, thereby linking the firms to their most relevant R&D institutions and consultants. The programme seems to be well adapted to the challenges met by industry in Northern Norway as well as the need to diversify and develop the industrial structure of the region. The funding of specific innovation projects combined with close follow-up has been highly supported by both industry and other initiatives inter-linked with the NT programme. The NT programme seems to fulfil a valuable role insofar as firms do not innovate in isolation but in active relation to other firms.

In conclusion, the main features of direct-support schemes are that the performance of SME support has improved in various fields: employment, productivity, management and workforce skills, profits, sales and market shares. But, some financial instruments have low additionality and there is a tendency to focus on R&D in isolation without sufficient integration with other forms of support necessary for the innovation process, particularly with respect to marketing and commercialization.

6.2.2 Results and Impacts of Technological Centres and Tools Fostering Technological Diffusion

A second barrier to SME innovation is the lack of technological information and know-how. This section studies the following technological centres and programmes fostering technological diffusion: in Austria the R&D-oriented technology centres SWP and FAZAT and the facility-oriented technology centres GTZ, TZL, TZI and TZS; the technology centres and service centres in Lombardia and Apulia; in Norway the Rush programme; and in Spain the four technological institutes AICE, AITEX, AIJU and INESCOP.[6]

Aims and target outputs of technological centres and tools fostering technological diffusion
The Software Park, Hagenberg (SWP) and the Research and Training Centre for Labour and Technology, Steyr (FAZAT) are two R&D-oriented technology centres located in Upper Austria. SWP is a technology and research centre for software development and related services and it was planned to comprise a mix of university institutes, companies and technical college courses. The FAZAT centre is located in one of the old industrial areas of Austria. Although the original plan was to establish an incubation centre, today the

project is a technology centre with a technical college. Its consulting activities deal with the attraction of firms to the region and with the support for start-ups and for existing companies regarding technological and organizational improvements to enable growth. In addition, the Austrian evaluation included four facility-oriented technology centres in Upper Austria. The Incubation and Technology Centre, Wels (GTZ) is an incubation centre which provides infrastructure. An important role of the technology centre is training and consulting services related to start-ups, the structural improvement of firms and the application for R&D support. The Technology Centre, Linz (TZL) and the Technology Centre, Innviertel (TZI) offer infrastructure and basic facility services. The latter offers some additional consulting services such as measurement technologies, quality management, application for R&D project support, and database and patent enquiries. The Technology Centre, Salzkammergut (TZS) is a Techno Park, still in the phase of foundation.

The goal of the Danish Technological Centres GTS is to provide SMEs with the possibility of accessing new knowledge and technologies through support in the development of projects related to technologies, entrepreneurial strategies and organization. GTS Institutes are to be seen as suppliers of knowledge to the Danish business community.

As regards other Technological and Service Centres, the intermediate structure operating to encourage innovation support for SMEs is the Tecnopolis/CSATA/Novus Ortus in Apulia. It represents a model of a technological park based on supply side, starting from the existence of academic competences in computer science and engineering at the University of Bari. The centre provides research and technological transfer, technological services in the field of informatics, training activities, services supporting innovation and location services.

In Norway RUSH aims at encouraging the college staff to enter into contract and development work for industry and to strengthen relations between institutions of higher learning and industry. Moreover, the idea is to indirectly enhance development of the business community and value-added production.

The ultimate goal of the four technological institutes selected in Spain (AICE, AITEX, INESCOP and AIJU) was to encourage competitiveness in SMEs in the ceramic, textile, footwear and toy sectors located in the area. Also, these instruments have two intermediate goals: modernization of the industrial fabric and diversification of production in the firms. To reach these objectives, services such as training, information, technical and technological advice, technological transfer and R&D projects were provided.

Most technological centres and diffusion of innovation programmes show positive trends in terms of number of clients, users, services rendered and awareness. For instance, the volume of RUSH activity, which includes registered industrial contacts, contracts for various services and the economic value in turnover is showing a positive development. Also there have been positive trends in the rapid mobilization of college staff and the persistent increase in staff involved in the four colleges participating in the programme. In Spain from 1989 to 1996, both firms associated with the four Technological Institutes and the clients of their services increased considerably, as did the volume of services rendered by the four centres. As regards GTS in Denmark, a market analysis carried out in 1997 indicated an awareness level of 55 per cent, an increase of over 46 per cent on 1996. In Italy most of the surveyed firms (63.3 per cent) have made contacts with at least one centre.

In Austria, the facility-oriented technology centre GTZ stands out. In its role as an incubation centre, the number of new firms has exceeded expectations. Since the early nineties the R&D-oriented technology centre SWP has also been continuously expanding. Regarding the use of services of the Upper Austrian technology centres, most of the surveyed firms are located in the centres. As far as external firms are concerned the R&D-oriented centres are less frequently contacted than the facility-oriented ones. However, the use of all technology centres is rather low: only 19 per cent of all SMEs use them. Within the set of SMEs with innovation support the services of technology centres are less frequently used than direct support programmes (35 per cent versus 90 per cent).

The general level of satisfaction with the support tools seems to be high, particularly in the cases of Spain and Lombardia. The Spanish technological institutes show a high degree of effectivity in adapting to innovation support needs as expressed by SME entrepreneurs in the four sectors studied. Eighty-two per cent of them point out that services offered by the Institutes adapted to their specific innovation needs. In general, the ensemble of services rendered by the Institutes conform to the firms' expectations (in 87.7 per cent of the cases) and the level of satisfaction achieved is good and very good for 85 per cent of the firms. In Lombardia the large majority (85.9 per cent) of firms contacting technology transfer or service centres are satisfied with the help obtained and define their degree of satisfaction at a good level. In Apulia a percentage of 63.3 per cent of the firms contacting at least one centre declared they were helped by the centre and defined the degree of satisfaction as somewhat lower: for 37.5 per cent of the firms it was sufficient, for 33.3 per cent it was good and 25 per cent considered the satisfaction level low.

In general, firms using technological centres and diffusion of innovation schemes are predominantly SMEs of traditional and service sectors. The primary target group of GTS enterprises is defined as SME and most of them are in the size class of 10 to 49. It is highly focused on SMEs, which is justified by the predominance of SMEs in the region. The same patterns arise in Italy and Austria, where the technology centres are most often used by the smallest SMEs (respectively less than 20 and 10 employees). The service sector is also important in the case of Italy, due to the increase of SME demand for services over the last years, especially in the area of computer science and robotics. In Austria service companies are the most frequent users of technology centres and only a few firms located in such centres are not service firms. Regarding industry differences SWP firms belong to the software and industrial mathematics fields; automation and telematics dominate the activities of FAZAT. In most facility-oriented centres software and data-processing firms are predominant.

The four technological institutes selected in Spain have their principal clients in the areas where they are located and target SME in the footwear industry (INESCOP), in ceramic activities (AICE), in the textile sector (AITEX) and in toys (AIJU). Although the Technological Institutes concentrate their provision of services in the Region of Valencia, in the last three years, they have had to search for new markets, catering to firms in the rest of Spain and abroad. Likewise, the institutes have had to work not only with SMEs but also with larger firms.[7] Nevertheless, most of their clients are SMEs located in the same specialized areas as the technological institutes. In the case of RUSH the regional emphasis is also important when SMEs in regional clusters form the main target group. The SME profile of RUSH activities is firms with less than 100 employees. RUSH's main target groups are non-R&D-intensive SMEs and parts of the institutional infrastructure (national R&D institutions and state colleges).

Effects of technological centres and tools fostering technological diffusion on client firms
Concerning SME product and process innovations, policy tools included in this section seem to have a limited impact, although in many cases they contribute to the improvement of existing products and processes among supported SMEs. In Spain approximately 58 per cent of the firms indicate that services provided by the Technological Institutes have helped them innovate. Promotion of innovation was particularly high in the case of AICE, both in creation of new products and in the improvement of existing products. All Austrian instruments seem to have a positive impact on product innovations. However, product innovations are less frequent in the case of SMEs supported by technology centres. Inducing process innovations is not the most influential area of these tools either, but, in the case of firms supported by technology centres, most process innovations are adoptions of existing technologies. An important effect of these technology centres may be found in the stimulation of a continuous or intensified innovation process within client firms. However, the survey shows that only less than

half of the firms indicated this longer-lasting effect. Improvement in this respect is certainly possible.

Additionality aspects have an important role among firms supported by technological centres and tools fostering diffusion of innovation. In Austria the most frequent answer as to the need of SMEs to use support instruments in order to carry out their innovation activities is that the innovation project or activity could not have been finished without the support. Technology centres are more often necessary to realize innovation projects than direct support. Services provided by Spanish Technological Institutes affect innovation within the firms in 58 per cent of the cases: services contribute to introducing product innovations in 39 per cent of the clients, improving existing products (31 per cent), introducing process innovations (24 per cent) and improving existing processes (18 per cent). Regarding instruments that foster diffusion of innovation, the Norwegian tool, RUSH, has a high degree of additionality. However, support tools in Denmark are an exception. There is no sign of any relationship between firms' investments in innovation and their use of support schemes. Thus, it is difficult to find any significant relationship between firms' use of GTS and their innovative activity.

The most important positive effects of the technological centres are those related to information, consulting, training and technical and technological services. In Spain these services provided by the technological institutes have contributed to improvement of product quality, level of skills of the work force and competitiveness in the client firms. In the case of AICE, their services have also facilitated the modernization of the productive process, improvement in productivity and the diversification of production in firms of the ceramics sector. In textiles, the use of AITEX's services has facilitated modernization of the productive process, an increase in sales and greater access to foreign markets. In the toy sector, AIJU's services have influenced sales positively and in the footwear sector, firms co-operate more efficiently thanks to the services of INESCOP. In Austria the provision of technical know-how, technical services and infrastructure contribute to increasing productivity level and growth. The supported innovation projects or activities have led to an improvement of productivity in many firms. Forty-one per cent of the firms indicated an increase in labour productivity, 32 per cent in capital productivity. The creation of new jobs is also encouraged by the technology centres (indicated by 46 per cent of firms). But all these effects are less frequent here than in the case of direct support.

Long-term effects and impacts of technological centres and tools fostering technological diffusion

Technological centres allow one to argue in favour of public support policies based on the notion of impact, that is, beneficial effects outside the firms receiving support. Thus co-operation with other firms and knowledge institutions and co-ordination with other regional, national and international programmes and centres, are considered in this section.

In Spain one of the characteristics of the sectors studied is the interaction and interdependence existing among the firms. Final consumer goods producers tend to outsource, relying on a large number of firms to execute the various stages of the production process. As a result, when the technological institutes enter into a service relationship with some of these firms, they establish at least indirect relations with many other firms in the sector that benefit from the support. Co-operation with other firms is a quite frequent consequence of innovation support in the case of Austrian firms using the services of technology centres. As to the types of partners, continuous co-operation with research institutions is less frequent (23 per cent) than with other firms (46 per cent). Spillover effects on other firms, most often in the form of improved products or services, were frequently claimed too.

One of the strengths of technological centres is their connections to other international centres of this type and the knowledge transferred to firms through them. In Spain it is important to recognize the role of the Institutes in diffusing international, national and regional information among the firms of the local economy. For instance the SMEs belonging to the sectors targeted indicate that they are highly informed of the various innovation support programmes existing in local, regional, national and European areas. Almost 93 per cent of the firms' entrepreneurs say that they were adequately informed on some of these programmes. Also in Denmark the GTS often joins forces with other international, national and regional programmes and players. It is thus a key to the strategy of GTS to integrate resources from different regional, national and EU partners in order to meet the needs articulated in the region. The Star project is prescribed as a good example of an attempt to integrate regional actors and mobilize resources in the regions to the benefit of regional business development. Activities for fostering co-operation programmes with firms belonging to other regions, the creation of an interface between research institutions and needs of local firms, represent the main areas of intervention of the Italian technological centres.

In the case of instruments fostering the diffusion of innovation, the Norwegian RUSH has tried to improve co-ordination with other programmes and support systems, but only the college of Vestfold could initiate a strategy of scale, aimed at paving the way for increasing the level of external activity in the region. The other colleges had more difficult contexts to handle. We have already indicated policy learning and changes in the support system as impacts of the programme. However, the direct impact of RUSH may be considered to be minor.

RUSH, as well as GTS and the Italian and Spanish technological institutes, are suited to regional clusters because they focus on encouraging collaboration among firms, cluster of firms and regional R&D milieus. In these cases, especially in Spain and Denmark, the diffusion of information, knowledge and innovations between the instruments and their client firms is determined by the integration of these centres in the productive and social fabric through proximity and the frequency of contacts between firms and the Institutes. The Institutes' strong points, then, are their proximity to the firms. Due to their feeling for the technological and collaborative patterns within the area, the interaction with local firms with different positions in the regional production system as well as with other supporting institutions, actors, regional development agencies, the building of social capital around specific areas of competence and repeated exchange and adaptation processes lead to a learning process between the network of clients and the instruments.

In conclusion, the main features of the technological centres and diffusion of technological knowledge schemes are the following. The technological centres have a limited impact as far as SME product and process innovations are concerned, although in many cases they contribute to the improvement of existing products and processes among supported SMEs. Additionality aspects have an important role among firms supported by technological centres and tools fostering diffusion of innovation. Also, the integration of the technological centres in the productive and social fabric foster the diffusion of information, knowledge and innovations through proximity and the frequency of contacts between firms and these centres.

6.2.3   Results and Impacts of Policies Based on Innovation Management and Brokers

Policies based on innovation management and brokers offer advice and consultancy to SMEs. The following tools are considered in this section: TICs from Denmark, KIC in Limburg, TEFT in Norway and BL, LVCs and LVBIC in the UK.

Aims and target outputs of policies based on innovation management and brokers

The Danish instruments, TICs, were created in order to modernize and renovate Danish SMEs by providing technological support and putting these firms in touch with other agents involved in innovative activities, as detailed in the preceding chapter. They comprise a network of regional based advisory centres that support industrial development in Denmark through activities in the 15 regional centres. Basically this is done as a broker between enterprises and a variety of technical and managerial consultants, experts, laboratories and research institutions.

In Limburg, KIC aims to strengthen relations of technological co-operation between the large firm OCÉ and the subcontracted supplying SMEs. This instrument focuses on inter-entrepreneurial co-operation among suppliers and clients. The final objective is to encourage these SMEs to learn from others belonging to the cluster existing around OCÉ and improve their technological level to be able to compete. It can be character-ized as cluster policy, focusing on learning how to innovate.

The two main objectives of TEFT in Norway are in the area of business development and infrastructure development. On the one hand, TEFT contributes to enhancing the capability of SMEs both in central and peripheral areas to initiate and carry out R&D projects. On the other hand TEFT helps R&D institutions to reorient themselves increasingly towards activities relevant to SMEs, in such a way that co-operation with smaller firms increases and that the knowledge base in these institutions becomes more accessible to all SMEs. Thus this tool tries to initiate behavioural changes in firms as well as in R&D institutions.

Some of the main objectives of the BL from the UK are to provide support services for firms, promote innovation and put firms into contact with suppliers of services and institutions that finance innovation activities. The original purpose of BL was to focus on supporting growth potential in established businesses rather than to focus on new or very young firms. BL provide advice on innovation and technology to innovative SMEs who show growth potential and aid them in identifying sources of innovation support.

The LVCs, studied in the UK, target SMEs and attempt to promote inno-vation, new technologies, new designs and entrepreneurial expansion in order to stimulate industrial and commercial growth in the area and the establishment of innovative and high-technology firms. The LVBIC, also in the United Kingdom, focuses on the provision of intensive support for innovative firms, with emphasis on technological opportunities, product innovations and commercialization of innovative ideas.

Various points related to the use of the policy tools can be identified. Between 1994 and 1998 TEFT exceeded its target number of firm visitations by a good margin, was slightly behind on its number of technology projects and had a higher average firm share regarding financing of projects that tar-geted. In the case of TICs in Denmark, 55 per cent of SMEs in the target group used the TIC system once or several times a year. Among the enter-prises joining the questionnaire for the SMEPOL study only 8 per cent have used the TIC, while 40 per cent knew about it and 12 per cent had been in touch with the TIC. In the case of LVCs, the pattern of client contacts for the 62 clients who participated in the survey was: 42 the prodesign centre, 32 the telematics support centre, 13 the technology transfer centre and 12 the teaching company scheme unit. The penetration of most of these initiatives measured by the number of firms assisted is typically small and awareness of the LVCs within the study region is low: LVCs 22 per cent and LVBIC 24 per cent (firms of the control group). Moreover the tools were used by fewer firms: LVCs 2 per cent, LVBIC 0 per cent. Nevertheless there are some positive experiences in the UK: BL was used by a significant minority of firms (28 per cent) and awareness was at 81 per cent.

Although in general the number of firms supported by these instruments based on innovation management and brokers is low, the level of satisfaction seems to be positive. In the case of TEFT, the overall result indicates a certain mismatch between the latent needs on the demand side and the reproducing mechanisms on the supply side, as 73 per cent of the firms report a good linkage between TEFT and the firms' business plan and 85 per cent of the firms report that they collaborated easily with the TEFT researchers. In the UK among the BL clients there is a high level of satisfaction with the support they received. A third of the firms rated the services as excellent and a further 49 per cent as good. Only 13 per cent described the services as average and 4 per cent as poor. As far as LVCs and LVBIC are concerned the results also indicated a high level of satisfaction. In the case of LVCs 60 per cent of clients rate the service as excellent or good, 13 per cent as average, and 10 per cent as poor or very poor. Most of the interviewees were very positive in their assessment of the help they had received from the LVBIC. Three quarters of those interviewed considered the quality of support overall to have been excellent or good.

The main target group of brokers and instruments based on innovation management are SMEs. As to size distribution, LVBIC assisted clients are predominantly micro-enterprises; in the case of the BL half of the innovating firms assisted were in the 10 to 49 size, over a third were in the one to nine size band and 14 per cent in the 50 to 250 band; the main target group of TEFT was SMEs in the range of 10 to 100 employees; and the target group of the TIC network is mostly SMEs between 0 and 200 employees.

For the most part, policy tools studied in this section concentrate their activity in the manufacturing and service sectors. For instance, TEFT-targeted industrial sectors were identified as the range of Norwegian industry, especially in sectors with low or medium R&D intensity. The traditional focus of TEFT in general is persistently on metal- and equipment-producing firms, sectors that account for approximately 33 per cent of Norwegian industry. In Denmark TIC's activities are mainly in manufacturing firms, with a primary focus on technology-oriented manufacturers and knowledge-based service enterprises.

Almost two thirds of innovative firms, which have sought help from BLs, are manufacturing rather than service firms. The manufacturing firms were distributed across a wide range of sectors, although most numerous were businesses involved in the manufacture of metal products, general and specialized engineering. The client firms, which were in services, were also spread across a wide range of activities, although firms in other business services and software consultancy were the most numerous. In addition a key aspect of the BL approach is to target SMEs with growth potential. In the case of LVCs the surveyed clients were spread across a wide range of sectors, with some bias towards service sectors. The main groupings of clients were in software, technical consultancy, wholesaling, food, clothing, and furniture. The kinds of services provided by the LVCs are more likely to be relevant to technology followers than to technology leaders.

Finally the profile of surveyed LVBIC clients is different. Firms cover a variety of activities, with many of them involving applications of advanced technology in robotics, telecommunications, electronic engineering and geo-informatics, but with few involving relatively simple technical ideas.

Effects of policies based on innovation management and brokers on client firms

Effects on innovation involve results related to product and process innovation as well as market innovations. The result varies among the policy instruments. For example, as mentioned above, in Denmark, only 12 per cent of the enterprises involved in product innovations have made use of support schemes. In the case of process innovations the percentage is lower (3 per cent).

On the contrary, firms supported by Norwegian TEFT report significant improvements in products and production technology, as well as increased R&D intensity and capability. The participation in TEFT had led to the following impacts: improvement in existing products (43 per cent), new products to the firm (35 per cent), improved production technology (40 per cent), increased R&D (41 per cent) and increased R&D capability (59 per cent). The effects relate mainly to increased knowledge about new technology or more general knowledge about managing development projects, increased learning ability, as well as changed attitudes and routines in firms and institutions.

Similar effects are inferred by BL in the United Kingdom. BL plays an important role in helping firms overcome barriers that SMEs were facing. Just under a third of firms said that the assistance they had received from BL had enabled them to overcome particular difficulties relating to making innovations. A substantial proportion of BL clients had been involved in making changes to their products and services over the 1993–98 period. BL clients introduced one or more completely new products or services (41 per cent), modified one or more of their products or services (49 per cent) and introduced a completely different range of products or services (22 per cent). Turning to changes in production processes, 79 per cent of BL clients made some change in terms of technology or organizational changes, 69 per cent of firms were involved in making technological changes and 43 per cent in making organizational changes. But the most important effect of BL appears in terms of helping client firms develop marketing innovations and exploit their innovative effort. BL clients were involved in opening up new geographical markets (57 per cent) over the period. The main development was through exporting, with 46 per cent of the firms opening up new international markets. In addition 60 per cent of the BL clients have developed new types of customers in the period and 64 per cent have introduced a new method for promoting their products and services. The comparatively high level of market development activity among BL clients should not be surprising in view of the fact that half of them received assistance from BL in the area of marketing, indicating that a high proportion were seeking to expand the markets for their products/services. This is an area where BL services would appear to be beneficial to client firms.

In Limburg KIC induced a more integrated approach to the internal innovation process, as innovation became part of the overall business strategy for most of the interviewed firms.

Additionality aspects also vary among tools based on innovation management and brokers. On the one hand, some instruments, such as TIC and BL have a low degree of additionality. In the first case in Denmark more than 75 per cent of the enterprises found that support has no part in the initiation of the product innovation activity. In the case of process innovations the pattern is even more extreme, namely more than 90 per cent of the enterprises. It is difficult to find any significant relationship between firms' use of support instruments and their innovative activity. In the case of BL in the UK only 12 per cent of the firms considered that they would not have gone ahead with the improvements. Thus, 88 per cent of the firms would have gone ahead with the project in some form or another. It appears that BL intervention makes relatively little difference as to whether firms go ahead with innovations or not as most interviewed SMEs consider that they would have gone ahead with the project, possibly using other sources of external assistance.

On the other hand, in some cases there is a higher level of additionality in firms that would not have gone ahead without help. For example TEFT showed a high degree of additionality. Also most client firms viewed the LVBIC as a unique and important resource for innovators and indicated that their projects would probably not have gone ahead without this support. In cases where innovation support involves finance and/or new venture creation (such as LVBIC), there is clearer evidence of projects proceeding with support that would not have done so without it. In Limburg KIC has helped to transform SMEs from 'jobber' to 'co-maker' and 'main supplier'. In some cases it has speeded up an already existing development and for others it has initiated it.

BL is fulfilling their objectives since, in terms of both changes in sales turnover and employment over the period 1993–98, the surveyed firms were growing, and in some cases rapidly. The median increase in sales turnover was 66 per cent and 69 per cent of the firms increased their employment. In terms of profitability, 78 per cent of BL clients made a pre-tax profit during 1997/98. Overall 41 per cent of the client firms surveyed were able to point to ways in which the performance of the business had benefited from BL assistance. In fact, 16 per cent of firms considered that the assistance had had a positive impact on sales turnover growth, employment generation and profitability, and a further 11 per cent thought that there had been a positive effect on at least two of these aspects. Just under a third of the firms considered that the assistance had helped them protect existing jobs within the firm, and a quarter claimed to have been able to create new jobs as a result. BL assistance does appear to be having a beneficial effect on employment in at least half of client firms. A quarter of firms overall thought that the assistance had helped them to increase sales turnover and/or improve their profitability. In addition 54 per cent of the interviewed BL clients considered that they had benefited in other ways from the help that they had received. Of these, a third considered that they had benefited in terms of assistance with the overall development of the business, and a further quarter considered that it had helped them to identify key issues affecting the future development of the business.

Long-term effects and impacts of policies based on innovation management and brokers

The degree to which the supported firms engage in new procurement of R&D services is an indicator of impact. Firms which enter into continuing relations with the same R&D institutions also enter into significantly larger projects. This is the case of TEFT in Norway that creates the foundations both for further external participation as well as increased R&D intensity in the firms. From the point of view of research engaged through TEFT projects, 68 per cent of them report that contact with the respective firms persisted beyond completion of the TEFT project and 48 per cent indicated that this concerned the planning of a new project. In sum, the results indicate that the general model of TEFT increases R&D and continuing demand for R&D services and works reasonably effectively, producing continued relations between the two parties.

Other possible impacts of policy instruments on the regions' innovation capability are by means of their ability to encourage interactive learning through co-operation between firms, between knowledge institutions and between SMEs and external organizations and programmes.

Regarding contacts between firms LVBIC is a successful example. Clients located at the incubation units of LVBIC spoke of the value to them of being in close proximity with other innovative small businesses. The key point here is that clients know by experience that physical proximity to other innovative businesses offers the possibility of new business opportunities, facilitates exchange of knowledge and is conducive to learning.

Concerning links with knowledge institutions and between SMEs and external organizations and programmes, most of the instruments play an important role. In Denmark, TIC Vejle finds that they are linked to the industrial development strategy provided by the region. TIC Vejle maintains ties to the national TIC network, partly because these links are essential to foster new programmes and tools. Norwegian TEFT helps link SMEs to other relevant policy tools or R&D institutions, thus playing a co-ordinating role in the regional support system. The LVBIC appears to be important as a network broker for clients who would not otherwise have the time and resources to develop their own contacts. It is valued as a node for accessing other networks, including other sources of specialist expertise such as universities and other government and EU support programmes. For instance the LVBIC facilitates access to government support schemes, notably SMART, and a number of clients had been directed to sources of advice within universities. BL might be seen as contributing to the development of a support infrastructure that links a public sector initiative with market-based private sector consultancy services, which is conducive to innovation in SMEs. Three quarters of the firms were directed to other sources of specialist expertise by the BL advisers. BL also helps SMEs engage with national and global information and learning networks as well as local ones in order to increase firms' competitiveness in the global marketplace.

Nevertheless in the UK, while there is some evidence of policy measures contributing to increased network activity, the overall impact is limited by the small scale of all but the BL. The key role of the instrument with respect to the innovative capability of the region's SMEs is to help increase their market orientation and the contribution to increase the use of external assistance and consultancy by SMEs.

The most successful experience among tools based on innovation management and brokers as far as long-term effects and impacts are concerned is KIC in Limburg. Furthermore it is especially relevant to the region, because there are a lot of industrial SMEs which lack codifying skills and Limburg also lacks a sound public R&D infrastructure. By creating co-operative structures, which promote interactive learning within the region, SMEs have the opportunity to prove their competence in co-development and engineering and to learn from other firms involved in the cluster. Thus regional SMEs have the chance to upgrade themselves.

In conclusion, policies based on innovation management and brokers are characterized by the positive level of satisfaction of SMEs with these policy tools, although the number of firms supported is low. There should be a trend towards the creation and use of these kinds of instruments. Concerning links with knowledge institutions and between SMEs and external organizations and programmes, most of the instruments play an important role.

6.2.4 Results and impacts of mobility schemes

This section includes the policy tools related to human resource barriers for SME innovation. The mobility schemes analysed are: FIRST and RIT from Wallonia and KIM from Limburg.[8]

Aims and target outputs of mobility schemes
FIRST from Wallonia aims to strengthen the scientific and technological potential of firms. It enables companies to gather knowledge that will be most relevant to engaging in innovation with a high degree of novelty. It was created with the goal of increasing firms' scientific and technological potential by contracting young researchers who work part-time in the firm and part-time in a research laboratory and whose wages are funded by FIRST over a period of two years.

The goal of RIT in Wallonia is to increase the capacity of human resources, thus encouraging the development of innovative projects. The RIT aims to help start innovation activity in firms through the strengthening of human resources. The mobility scheme KIM in Limburg, is designed to promote internal processes for innovation and mobility of human resources as well as impulse technological transfer and diffusion of technical and management skills from institutions of higher education to firms. In order to achieve these goals, KIM funds the initial stages of technological innovation in SMEs through reduction of labour costs or contracting a recent university graduate for one year.

FIRST is by order of importance the second most frequently used instrument in Wallonia. Statistics on the use of the FIRST instrument indicates that it has more than doubled the number of dossiers initiated from 1995 to 1997. On the other hand RIT has a low level of use throughout the three years under consideration. In Limburg, 44 per cent of all the participants already knew of KIM before their application.

The chemical sector stands out in terms of FIRST support consumption, followed by other high-tech sectors such as electronics and the business service sector. The more low-tech or traditional sectors like textiles, wood or leather are few. FIRST is highly relevant for both high-tech and low-tech firms with a long-term innovation strategy that intend to build a knowledge base within their company as a source for future innovation activity. In practice, the instrument predominantly reaches firms that are most advanced in innovation. It is predominantly used by SMEs, although SMEs were not specifically targeted.

RIT specifically targets SMEs. Each sector has at least accessed one RIT over the three years, including sectors that are considered as low-tech or traditional. The instrument reaches the type of firms that it implicitly targets; non-innovators but with innovation potential. It is highly relevant to non-innovating or newly-innovating SMEs. Two thirds of firms participating in KIM had less than 20 employees. The average size in the National Schemes is 16 employees. However, the size criteria for the KIM–Limburg scheme is less strict than in the National scheme and the average size of the participating firms is 19 employees. In addition the KIM scheme seems very relevant to participating firms, since in 20 of the 30 firms there was one employee holding a university degree and in the other 10 firms there were none.

Effects of mobility schemes on client firms

As far as additionality aspects and effects on innovation and business performance are concerned the mobility schemes seem to have positive results. In Limburg many of firms' innovation activities would not have taken place without the KIM project. An indication of the effect of KIM on the innovative behaviour of the participating firms is based on the following perception: 30 per cent of the firms claim that the innovation would not have taken place without the KIM project; 50 per cent of the participants said the project had speeded up or extended the innovation; 20 per cent claimed the project had not yet had a clear influence. The most successful effect of KIM is that it has lowered the barrier to employment of persons holding university degrees, since a large majority claims it would not have hired a graduate to fulfil the innovation task and 79 per cent of the projects did result in an extension of the employment contract after the labour-cost subsidy had stopped. The success of KIM is in lowering the barriers for small, knowledge-extensive firms to enter the higher segment of the labour market. Firms appeared to have a more positive attitude towards high-educated employees and towards innovation than before. Also the interviews indicate positive effects of the presence of the KIMer within the firm.

In Wallonia the main effects of FIRST relate to training and knowledge as well as access to external resources for R&D. Theoretically it enables the employment of high-quality researchers, which could otherwise not have been afforded by the company. This case constitutes an important additionality to the set of direct-support instruments in Wallonia. The RIT stimulates the employment of staff who are more highly qualified (often engineers) than existing staff within SMEs, and the firm would not have been able to employ such staff without support. The latter constitutes the main additionality of the scheme, from which most other benefits derive. The RIT has a potentially high behavioural impact on SMEs. The employment of an innovation and technology manager often marks the beginning of continuous innovation activity, through creation of an R&D unit and increased human resources. RIT clearly influences the decision of non-innovative SMEs, often in traditional sectors, to engage in innovation activity. For most beneficiaries, access to RIT has accelerated this decision. For the least innovative ones, the impact was strongest: strengthening of human resources, creation of an R&D unit, SME growth. Moreover a number of RIT beneficiaries displayed substantial growth in the 1980s, which corresponds approximately to the initiation of innovation activity. RIT is at the origin of this development: innovation activities opened up new market opportunities and generated employment.

Long-term effects and impacts of mobility schemes
FIRST's strong point is clearly its potential to generate impacts. The instrument encourages a long-term approach to innovation and enables acquisition of generic knowledge that can be applied to a variety of products as well as enable the development of products with a high degree of novelty. In the long run FIRST, more than any other instrument at work in Wallonia, could be the source of radical innovations. On the other hand RIT fits in well with the internal nature of the innovation process in Walloon firms. This increases relevance but limits change in behaviour with regard to extra-firm interaction. An important impact of the tools is that, generally, firms that have benefited from a RIT as their first innovation support have later on accessed other support instruments.

In Limburg KIM addresses the large category of firms which are driven by a learning-through-producing behaviour. Thus for several reasons KIM is especially relevant in the case of Limburg. A positive further impact within two regional KIM schemes is the creation of a network of knowledge carriers. The knowledge carriers (KIMers) in those regions organized meetings where they could discuss all kinds of subjects and exchange experiences from their individual projects.

Thus, FIRST encourages a long-term approach to innovation, RIT fits in well with the internal nature of the innovation process in Walloon firms and KIM addresses the large category of firms which are driven by a learning-through-producing behaviour.

6.3   WHY DO SOME POLICY TOOLS WORK BETTER THAN OTHERS?

By means of a benchmarking process of the four groups of policy tools analysed, this section points out the instruments from each group that have been most successful in alleviating barriers to innovation in SMEs. The selection of these instruments was effected in terms of their results in the area of aims and target outputs, use and level of satisfaction, target firms by size, sector and technological level, effects on client firms (innovation, additionality and business performance), long-term effects and impacts.

Two good practices of the SME innovation support policy designs are chosen and the strong point of each one is described in the area of key issues relative to policy and interactive learning, adaptability and flexibility, responsiveness of policy to firms of different types and different stages of development, subsidy element, participation of firms in the design and management of the tool, ability to identify needs of the firms, complementarity between policy instruments and network activity.

From the direct-support tools we selected the NT from Norway and the SMART from United Kingdom. NT is matching funds with other programmes and links the firms to their most relevant R&D institutions and consultants. Thus the strong point of this instrument is its complementarity between different policy tools. The NT programme seems to fulfil a valuable role insofar as firms do not innovate in isolation, but rather in active relations. SMART shows evidence of policy learning and adaptations aimed at addressing the financial problems of SMEs.

From the second group of policy tools we selected the Technological Institutes in Valencia (Spain) as one of the good experiences. Entrepreneurial participation (voice) in Institute management as well as cost for the use of its services (subsidies average 40 per cent of the cost of services provided) contributed to adjusting supply and demand for innovation support. As a consequence, competitiveness of area firms has improved. It is for this reason that cost and voice in the policy tools are so important. Except in cases of new industry and enterprise, the concept of cost to firms reinforces the idea that subsidies granted by the administration are a collective good and, as such, cannot be subject to opportunist acts of rent-seeking. Policies based on total gratuity of subsidies for innovation encourage misuse of public funds, while policies designed to provide services at a cost to firms will indirectly reveal the benefits of innovation to the firms. Likewise, entrepreneurial voice in the design of instruments of innovative action will tend to avoid policies that do not correspond to the industrial reality. Introducing measures such as entrepreneurial participation necessarily brings policies nearer to the existing socio-economic reality and also compensates lack of industrial and innovative maturity in a given territory. Among the second type of tools we also selected the GTS from Denmark. Like the Italian and Spanish technological centres, this instrument is flexible in the sense that it is suited to regional clusters since they focus on encouraging collaboration between firms, clusters of firms and regional R&D milieus. In these cases, the diffusion of information, knowledge and innovations between the instruments and their client firms is determined by the integration of these centres in the productive and social fabric through proximity and the frequency of contacts between firms and the institutes.

Among policies based on innovation management and brokers, we selected the BL from the United Kingdom and KIC in Limburg. BL plays an important role in helping firms overcome barriers that SMEs were facing and the main effect appears in terms of helping client firms to develop marketing innovations and to exploit their innovative effort. Thus its responsiveness to firms that need to exploit the results of their innovations is the strong point of this policy tool. In addition, it contributes to increased network activity. The other good practice is KIC in Limburg, whose objective is to strengthen relations of technological co-operation between the large firm OCÉ and the subcontracted supplying SMEs. By creating co-operative structures, which promote interactive learning within the region, SMEs are able to show their competence in co-development and engineering and to learn from other firms involved in the cluster.

The policy tools selected as the good practices among the mobility schemes are KIM, also from Limburg, and RIT from Wallonia. The success of KIM lies in its ability to identify SME needs, lowering the barriers for small, knowledge-extensive firms to enter the higher segment of the labour market. The RIT is specifically targeted at SMEs and it reaches the types of firms that it implicitly targets: non-innovators with innovation potential. The responsiveness to non-innovating firms with innovation potential, often in traditional sectors, is the strong point of this instrument since it allows firms to engage in innovation activity through the reinforcement of their human resources.

7. Coherence of innovation policy instruments

Poul Rind Christensen, Andreas Peter Cornett and Kristian Philipsen

7.1 INTRODUCTION

In the previous chapters as well as in the national studies, it has been stressed that the formation of an integrated or coherent innovation policy system is a necessary condition in support of a successful regional innovation system.

In Chapter 3 the regional barriers to innovation were emphasized. It was stressed that organizational thinness as well as a fragmented regional innovation support system and – at the other extreme – a tight local support structure institutionalized in a regional area all may lead to 'excessive' barriers for innovation among SMEs.

In Chapter 4, it has been stressed that the population of SMEs is heterogeneous regarding their innovative activity in several respects of the word. Their needs as well as their contextual situation differ strongly. Therefore, it is concluded, lateral communication is needed in order to foster dynamic and adjustable support instruments.

The conclusions of Chapter 5 stress the need for interactive policies that will ensure 'client-focused services' and help to implement interface structures between industry and research institutions. These points call attention to a critical puzzle, namely the extent to which bureaucratic rationality in the organization of innovation support programmes is a barrier by itself to the creation of policies responsive to the dynamic situation and changing needs of SMEs which they intend to sustain.

Based on the propositions and conclusions drawn from previous chapters, this chapter takes on the task of reflecting on the ambiguous notion of coherence. Tentatively, the essence of the notion is that a coherent innovation policy provides solutions on specific issues in an integrated way and that customers or the target group perceive them as coherent. Section 7.2 will elaborate on this statement.

Policy-makers and planners have tried to implement this ambition through steering, support measures, the development of logical frameworks, routines and incentives. An internal logic or even lock-in of the support system is the typical byproduct evolving from these efforts. In summary, these efforts are seen as attempts to create supply-side coherence of the system.

However, traditional bureaucratic systems focusing on top-down more than lateral relationships rely on various measures of hierarchic govern-ance, often with a lack of flexibility. On the other hand, while flexible response to changing external dynamics and stimuli is the force of interac-tive innovation support systems, the lack of co-ordination and supply-side coherence of the policy implementation represents potential weaknesses following from the effort to promote responsiveness and the decentraliza-tion of policy execution, for example.

Fundamentally, there is a basic contradiction between the aim of respon-siveness to diversified regional needs as well as to dynamic changes in clients' needs and conditions on one side and the aim to build co-ordinated programmes for the sake of scope and administrative simplicity and trans-parency on the other side.

Based on this proposition, this chapter has several purposes. In section 7.2 a detailed discussion on the concept of coherence is provided as viewed in terms of responsiveness and co-ordination. Coherence is considered important in support policy programming and the evaluation studies of policy programmes. Therefore it is important to take into account differ-ences in the planning tradition upon which the concept is built and to con-sider to what extent we have to 're-invent' the concept when the perspectives in support programming and the perspectives of evaluation are changing.

In section 7.3, focus is on specific aspects of coherence in programmes revealed in the national studies of the SMEPOL project. The regional per-spective taken in the study has the implication that regional diversity will be confronted with national perspectives in policy programming. An anal-ysis of demand-side and supply-side aspects of innovation policy becomes crucial in this respect. This study has a focus on SMEs and their respon-siveness to innovation support programmes. This leaves us with the basic question of efficient delivery systems and thus the issue of proximity to SME clients.

The country studies carried out have suggested that the dynamics of post-Fordist society lead to a number of difficulties for traditional types of innovation management and innovation policies. Basically, an interactive support system is seen as a better fit than the linear approach, which dom-inates in most of the countries studied.

Therefore, in section 7.4 a dynamic setting is briefly discussed. Based on this discussion an integrated view on coherence in the policy process is put forward. The provision of a few theoretical perspectives completes this section.

Chapter 7 is concluded with a discussion of how to conceptualize coherence in policy programming so that it matches the industrial dynamics of the transforming society.

7.2   ON THE CONCEPT OF COHERENCE

In a regional innovation system perspective, at least two aspects of coherence seem to be important. Coherence from an administrative or a supply-side point of view denotes that we have an unequivocal relationship between means and goals. With regard to implementation this notion is similar to traditional concepts of societal planning. This is shown in section 7.2.1 below.

Seen from a client or demand-side perspective, the concept becomes muddier since the individual firms participating in innovative projects will interpret coherence in idiosyncratic ways. The projects are defined by the specific needs generated from current activities and the situation, which shape their needs more than the generic characteristics of the system. This schism is important when examining regional innovation policy from the demand-side perspective.

In Table 7.1, these two aspects of coherence are sketched out together with other central aspects of innovation policy in Western Europe: the issue of decentralized vs. centralized policy frameworks, in this case, the top-down/bottom-up distinction. The table will be used as an initial classification of innovation policy processes. In a second step, the intersection of the supply- and demand-side dimension with the notion of internal and external coherence will be added. (See Table 7.2.)

Although the bottom-up approach tends to be fuelled by the proximity to and thus awareness of specific client needs, bottom-up processes may easily be formed in a closed circuit of vested interests and support paradigms embedded within different regional actors of the innovation support system. An example of this type of lock-in is provided in Chapter 3, in the case of Lee Valley and Valencia. On the other hand the top-down processes may in practice evolve into partiality based on vested interests and differences in outlook held by different central planning agents.

To sum up, the bottom-up approach focuses on regional initiatives and the needs of the business community, while the top-down approach deals with the implementation of national programmes and measures in a regional setting.

7.2.1   The roots of the concept

Although coherence is one of the key concepts in the evaluation of policy programmes and support schemes, it is seldom referred to in conceptual

Table 7.1   General aspects of coherence: an initial classification of the innovation policy process

	Top-down	Bottom-up
Coherence from a demand-side perspective	Scope and integration in the delivery and implementation of programmes	Effectiveness in customizing programmes to clients' needs. Raising clients' awareness of programmes
Coherence from a supply-side perspective	Effectiveness in reaching objectives. Efficiency in co-ordination of programmes	Framing and tailoring of programme packages matching diversified needs

terms. Most often the concept is defined in negative terms, stressing the lack of coherence. In its generic form the concept is defined in the Encyclopedia Britannica (1999) as 'systematic or logical connection or consistency' or 'integration of diverse elements, relationships, or values'. Basically, coherence thus means that we have links between different elements, together constituting a holistic unit. The functioning of the whole system depends on the system being coherent and the elements working together properly. This concept of coherence can be linked to production systems, policies or business strategies, but does not necessarily require formalized links (for example, demand-side coherence is established if the users perceive the policy measures as a comprehensive offer). In a study on corporate coherence Teece et al. (1994) state that a firm exhibits coherence when:

> . . . its lines of business are related, in the sense that there are certain technological and market characteristics common to each. A firm's coherence increases as the number of common technological and market characteristics found in each product line increases. Coherence is a measure of relatedness. A corporation fails to exhibit coherence when common characteristics are allocated randomly across a firm's lines of business. (Teece et al. 1994, 4)

Although the term coherence is referred to in empirical studies and evaluations on industrial policies, the concept of coherence is difficult to trace in the literature of social sciences not to mention literature inside the tradition of planning. It is thus unclear in which planning tradition the concept was born. The literature on strategic planning revealed no references on the concept. There is also no presence of this concept in reviews and directories of public planning. The traditional notion of coherence is, at the outset, difficult to associate with feedback processes inherent in interactive or decentralized concepts of programming processes. Therefore changes in the theoretical perspective about the way that we conceive the innovation activity also leads to new ways of conceptualizing coherence.

The essence of the concept is that a coherent innovation policy provides solutions on specific issues in an integrated way (supply-side coherence) and that 'customers' or the target groups perceive them as coherent (demand-side coherence). In this respect the concept of coherence is important as far as it is conceived to be critical in respect of efficient and successful innovation policy programming. On the other hand coherence may be conceptualized in two other dimensions, namely those of internal coherence and external coherence. The former is associated with the integration and scope of individual programmes in isolation, while the latter is associated with the cross-sectional integration of different programmes aiming at the same target group. The conceptual relations are stipulated in Figure 7.2 below.

Table 7.2   The intersection of two dimensions of coherence

	Internal coherence	**External coherence**
Supply-side coherence	Coherence inside individual programmes (or support schemes)	Coherence across programmes of ministries and relevant planning bodies
Demand-side coherence	Scope and integration of individual programmes is appreciated by targeted actors	Programmes are found by the target groups to be well co-ordinated and tailored to current needs and context

The dilemma inherent in the supply side and the demand side of coherence is that although a programme may be found as being internally as well as externally well-integrated with other programmes and is seen to be efficient in terms of achieving the balance between means and ends, it also has to match the needs perceived by the targeted client group. This does not merely imply that they have to know about it and value it. It also implies that delivery processes are executed in ways perceived as efficient by SMEs.

In terms of a full scale – and idealized – evaluation perspective, a successful programme is associated with an outcome that fulfils all aspects sketched out in Table 7.2. If the demand-side aspect of coherence is not sufficiently met then policy implementation will fail. This may, for example, be caused by a lack of awareness of clients needs, their contextual constraints or simply by a sole supply-side perspective dominating the efforts of the support system. If supply-side coherence is weak, on the other hand, the innovation policy becomes vulnerable to organizational slack, misuse of resources due to a lack of programme co-ordination and competing programmes in the business development system, i.e. programmes focusing on issues other than innovation.

7.2.2 Conceptualization in a Dynamic Setting

Coherence so far has been discussed along the traditional line of arguments, namely as a matter of coherence of programmes, be it internal coherence of individual programmes or external coherence among related programmes. In this discussion the organizational coherence tends to be neglected or at least included implicitly. Organizational coherence of the support system does have a bearing of its own for the simple reason that in a temporal perspective, programmes change, while the organizational set-up tends to be stable.

In a dynamic setting, the ability to maintain coherence between the programme building and the organizing of programme implementation according to the changing needs of the target groups of the programme is the core essence of a responsive and interactive system of innovation policy. The organizational construct carrying programmes downstream to final users targeted is thus a key not only to the way clients are focused on, but also to the continuity in the delivery system of evolving support programmes.

Therefore a major emphasis should be placed on the conceptualizing of coherence in a programme as well as an organizational perspective. In the SMEPOL study, a coherent innovation policy system is defined as one in which individual programme elements can be organized and combined in a way to form a coherent tool for solving specific problems of innovation in a specific spatial context over time.

Focus is thus on policy processes and the relationships between instruments in a regional setting. Included in this conceptualization are not only the internal and external coherence of each of the policy programmes investigated, but also the organizational set-up, linking policy formulation with that of delivery and implementation. The aim is to reveal the change of perspective when a regional innovation system perspective is taken and when a dynamic – learning – perspective is introduced in the analysis of the innovative activity of small and medium-sized enterprises. This approach is sketched out in Figure 7.1.

The aim of the figure is to illustrate that in the traditional 'linear' notion, evaluations and discussions on policy programming are taking place predominantly in the organizational space of overall programming. In this setting, discussion is on coherence of programmes (the bureaucratic supply-side perspective). Organizational coherence may be low, that is when the main focus is with internal coherence of individual programmes. On the other hand organizational coherence may also be high when individual programmes are co-ordinated across the functional fields of different policy bodies. In this setting organizational coherence gains momentum indicating a growing emphasis placed on external coherence at the national level of policy programming.

Spatial coherence is reached when emphasis shifts to the regional dimension and thus the problems of co-ordination of the innovation policy delivery system are added. Co-ordinated action at the regional level is, by and large, conditioned by a co-ordinated delivery organization at the regional level. As the previous chapters have shown, the regions in focus in this study vary greatly with respect to their organizational capacity to implement, not to mention their capacity to initiate, innovative policy programmes.

The figure thus illustrates two fundamental dimensions in the pathway to a coherent innovation support system focusing on SMEs in a regional context (illustrated by the diagonal arrow in Figure 7.1). On one axis is the organizational coherence producing and implementing a co-ordinated or even integrated package of innovation support programmes. Dialogue and mutual adaptations are in this dimension, which is of importance to horizontal learning processes. On the other axis there is the regional (spatial) coherence comprising the organizational backbone at the regional level, which is the key to a vertical dialogue concerning programme deliveries responsive to regional needs.

While focus has been with the horizontal organizational coherence for a number of years, much less emphasis seems to be devoted to the vertical dimension of spatial coherence. However, if one agrees that focus ought to be with those SMEs and regions in weak positions in the current transformation to a knowledge-based economy, then it seems evident that the spatial coherence comprising the organizational set-up at the regional level, as well as the organization of the delivery system focusing on servicing clients at the regional level, are major bottlenecks to the promotion of innovative incentives and activities among SMEs.

When the regional dimension is included in the evaluation of innovation support programmes, the number of actors involved expands. Their positions in the field of overall programming tend to become highly differentiated, as does their frame of reference. Therefore, their perspectives on, and interpretation of, how the innovation support programmes function tend to differ strongly.

It is in this perspective of fractioned views and perspectives that vertical as well as horizontal learning cycles among actors with different perspectives, positions and outlooks in the system become of vital importance to the production of coherence. The establishment of learning cycles between the regional level and the national level in the organizational system carrying the innovation policy programming as well as between the business community and the policy-making community is thus seen as a fundamental condition for a working innovation system in most national studies.

Contradictions – or 'trade-offs' – between supply-side organizational coherence, and the need for rapid and flexible adaptations to a dynamic regional context and rapid changes in situational conditions, constitute barriers to a regional system of innovation targeted at meeting the needs of SMEs as well as the general business community of the region considered.

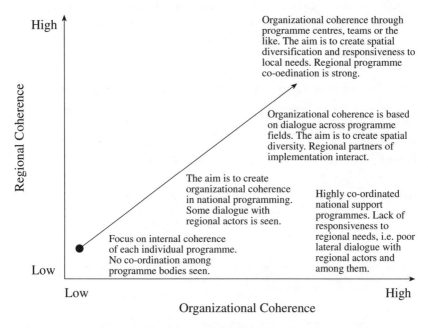

Figure 7.1   Organizational and regional coherence in an interactive perspective

7.2.3   Planning and Direction of the Policy Process

The top-down aspect of innovation policy and thus coherence has to be interpreted in the tradition of public planning and policy implementation. This means that innovation policies in this tradition do, to a large extent, rely on bureaucratic measures like rules, administrative routines and procedures, delegation and co-operative measures. At the same time it is a system of vested interests embedded within, say, ministries and similar fields of responsibility. The traditional system operates with a system of governance based on budgets, specific orders or the setting of targets. In Table 7.3 below the principal aspects are summarized.

Elements of this conceptualization can often be found in innovation systems, in particular within the linear tradition, in which the top-down perspective prevails. Direct governance, as outlined in Figure 7.4, is most often closely associated with the linear models of planning. On the other hand the 'interactive approach' is to be associated with a planning environment dominated by indirect governance (bottom-up).

In many respects the dimensions of the planning processes summarized in Table 7.3 can also be found in innovation policy systems. A hierarchical organization will provide coherent decision-making rules, form an effective bureaucracy using top-down approaches and predominately direct and specify measures.

Table 7.3   Systems of governance in the (public) planning system

Specific programmes	General programmes
Direct governance (Linear planning model top-down)	Orders, rules Targets, Programmes
Indirect governance (Interactive approach bottom-up)	Information, dialogue Budgets Guidelines Recruitment Procedures

Source: Based on Lundquist (1974).

In a network-based system, the patterns of incentives differ strongly from those of the hierarchy. Independent bodies and actors pursue their own objectives and they judge the programme and its implementation in light of their own agenda. Programming thus has to rely on collaborative ventures formed by mutual adaptation processes and co-ordinated action based on information exchange and dialogue, i.e. mainly indirect means of governance, substituting rules and direction with frameworks and incentives as indicated in the notion of 'Indirect governance' in Table 7.3. The nature of this policy system can be bottom-up, top-down, or, most often, a combination of the two.

7.2.4   Coherence in a Linear and Interactive Perspective

Internal coherence has its focus on the internal logic of the instruments and policies of innovation, meaning that means and action planned – and executed – are in support of the aims put forward. Internal coherence is therefore particularly in focus when discussing coherence within a linear perspective. Most of the concepts discussed in the country surveys deal with this notion, which is similar to traditional models of internal rationality used in mainstream planning traditions. Coherence, from an interactive perspective, is much more complex and it can at least be questioned whether or not it is possible or fruitful to make a strict delineation to distinguish between internal and external coherence in this case. The rationale of internal (programme) coherence can be illuminated as a hierarchical correspondence between different tiers in the policy programming process as illustrated in Figure 7.2 below. Basically, the core of the innovation policy system is set in between the framework formed by the policy system and the targeting of clients' needs.

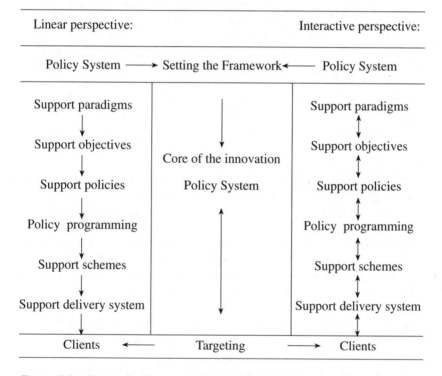

Figure 7.2   Internal coherence in a vertical perspective

In the linear model the downstream arrows are the dominant ones. From an interactive perspective both directions of the arrows are considered to be important. Depending on the nature of the programme they may be extended by feedback loops between all elements. Before turning to external coherence it is necessary to specify the character of the innovation policy clients. The business community and industry and in particular SMEs are the direct targets of the innovation support policy in this study. However, innovation policy also aims indirectly to improve regional growth potential and the development of, say, regional employment and social welfare benefits, through the creation of comprehensive regional innovation systems.

7.2.5 External Coherence

From the traditional linear perspective external coherence tends to be conceptualized as the horizontal co-ordination of related programmes embedded with different ministries and government bodies. The introduction of a regional perspective brings a vertical dimension into focus. This means the co-ordination of programmes and action, from international and national sources (the EU and national governments) through to regional and local bodies pursuing their own targets. These regional and local bodies often play a key role as the 'downstream' delivery system in proximity of the targeted SME clients.

The vertical dimension thus comes up in the form of organizational co-ordination among the various actors involved, from initial programming to final implementation and evaluation among the clients and target groups. When small firms are focused on, demand-side perceptions of coherence in the policy programme are naturally included.

Therefore, from an interactive perspective, external and internal coherence tends to become more and more amalgamated, since the interactive dialogue and way of co-ordinating the process will alter all features of the concepts simultaneously, including horizontal and vertical co-ordination. The contextual frame of reference may differ, as may the administrative tradition in which the notion of coherence is rooted. Therefore, a few key questions have to be addressed in the light of these differences in the national framework of reference. These questions are:

- How has the regional innovation support system evolved and how is vertical co-ordination between the levels brought about in the national studies?
- How does the view on coherence change when a dynamic perspective is introduced?
- To what extent has a tradition for learning processes become institutionalized?

These issues are addressed in the next section.

7.3. THE NOTION OF COHERENCE IN THE NATIONAL STUDIES

Coherence in innovation policy as stated in the previous sections depends on supply-side as well as demand-side features of the sector analysed. In particular, it is worth stressing that the outcome of a particular study has to be evaluated in terms of a temporal perspective and territorial space. A wide range of factors, internal to the system as well as external, determines the national systems of innovation.

In the context of the system of innovation, the institutional aspects are of importance, as stated by Acs and Varga (2002), in particular with regard to the role of the public sector:

> In the real world, the state and the public sector are rooted in national states, and national borders define their geographical sphere of influence. The focus on national systems reflects the fact that national economies differ regarding the structure of the production system and regarding the institutional set-up. (Acs and Varga 2002, 142)

Coherence is also affected directly when EU programmes are integrated into the national and regional system of innovation. This is not always conducive in terms of SME participation as this quote indicates:

> During its life some of the most burdensome regulations for small firms have passed through the Council for Ministers without so much as a peep from DG 23 . . . (Storey 1994, 266)

Large countries tend to have more heterogeneous systems of production than smaller countries, which results in less coherent national systems of innovation. Therefore, the build-up of regional innovation systems and thus also regional responsiveness in policy programming and action tend to gain importance. This may be seen as one of the reasons why analytical focus tends to narrow in on regional innovation systems. Regional innovation systems tend to be of a more integrated and coherent nature than national systems, implying also that the dimension of support policies has the option of a stronger coherence.

7.3.1 Perspectives on Coherence in the SMEPOL National Studies

The aim of this section is to provide an overview of the various concepts of coherence used in the survey of national innovation systems and programmes. Most surveys have adapted these concepts to the investigated programmes.

In the Austrian study coherence is understood as: '. . . the overall adequacy and effectiveness of the investigated support instruments in stimulating and improving the innovative performance of SMEs' (Kaufmann and Tödtling 1999). Internal coherence deals with the adequacy of the means to attain objectives; external coherence describes the integration into the whole support system.

The internal coherence of the direct support programmes investigated is quite satisfactory. However, in some cases, especially the RIP programme, there are inherent contradictions regarding the targets for innovation and employment/regional development. The technology centres are more difficult to assess in this respect. The R&D-oriented centres are successful regarding the stimulation of R&D and innovation. These are explicit objectives. The facility-oriented centres, on the contrary, are aiming more at incubation, an objective that is well achieved, but does not focus much on the degree of innovation or the intensity of R&D. This is not to be considered a lack of internal coherence as far as the centres are concerned. Such an inconsistency can only be argued for in the technology/innovation policy.

External coherence is also achieved as far as the existing direct support instruments are concerned. Their activities are well co-ordinated. Nevertheless, there are elements lacking which are necessary for a comprehensive innovation support system (e.g. venture capital, innovation management consultancy, focus on those firms which lack the capabilities to innovate). Referring to the technology centres the main weakness in external coherence is the lack of relationships with external firms. The centres are concentrated on the firms located in the centres, leading to a lack of regional embeddedness.

In the Norwegian study, internal coherence is conceptualized as the link between objectives, goals and targets of the policy instruments and the means to achieve these. The concept is internal to the programmes considered and seems to have strong similarities with the concepts of (bounded) rationality used in public administration.

Regarding external coherence, an interesting point of the Norwegian study is: 'A general conclusion from the evaluation of the three policy tools is that learning does take place and that this in a way has reduced the external coherence of these programmes (especially NT and TEFT) since they were implemented' (Isaksen et al. 1999, chapter 5 p. 178).

The notion of external coherence in the Norwegian study refers to the embeddedness of the individual policy instruments into the general support system, and the extent to which new ideas and results of the policy are built into the system. This seems to indicate a shift towards a more dynamic and interactive system of innovation. The price is a lower or a reduced level of external coherence of programmes in the traditional sense of the word.

In the terminology of this chapter this tendency represents a turn towards a more responsive typology of bottom-up processes and, probably, also a stronger demand-side driven system of innovation support. Overall, both external and internal coherence are seen to be satisfactory in the Norwegian case. TEFT and NT fulfil reasonable requirements to the level of internal coherence between means and goals of the programmes. Only RUSH has a rather mixed record, mainly due to the size of the programme and the limited means involved. With regard to external coherence RUSH meets the target set with regard to implementation, but it has a weak element of co-ordination built into the programme. The latter is often solved by on-site co-ordination at the regional level. TEFT as well as NT are seen to have achieved a satisfactory level of external coherence with the exception of some modest problems concerning the implementation of TEFT.

In the Dutch and Belgian programmes the record of coherence is seen as rather mixed. Internal coherence is mainly dealing with means and goals (objectives) of the programme being analysed. External coherence is a wider concept dealing with whether the programme is linked to other policy measures and instruments seen from the policy-makers' (supply-side) per-spective as well as the customers' (demand-side) perspective. Internal coherence (modest or poor in RIT) and external coherence and co-ordination are generally weak in all programmes considered.

The British study addresses two aspects of coherence. The notion of internal coherence is mainly discussed as a matter of co-ordination and comprehensiveness of targets and instruments:

> Nevertheless, since on balance each measure has a distinctive purpose and target group, with a degree of internal coherence, there is a degree of complementar-ity. (Smallbone et al. 1999, 218)

In the latter part of the quotation, external coherence between different programmes is seen as a key to good performance. The Business Link system in particular seems to provide tools for the creation of a coherent system of business services covering commercial as well as technical aspects, which, according to the summary of the UK report, leads to the following conclusive notions:

> The holistic approach of innovation adopted by BL means that SMEs which are seeking help with product development or particular technical problems may end up receiving assistance with the marketing of their products and service, ensuring that they reap the full commercial benefits from their innovations. (Smallbone et al. 1999, 5)

With regard to linking SMEs to other public innovation support pro-grammes and linking them to the Higher Education Sector, the BL system is less successful. But it does seem to be the case that the BL system in par-ticular is well received by its clients. This does leave us with an indication that BL-like systems serve well as local and regional deliverers of innova-tion assistance to SMEs.

So, in contrast to, for example, the Danish study, in which local and regional bodies tend to act autonomously and un-coordinated in their approach to client enterprises, the UK system seems to function in coher-ent ways at the regional level. The Business Link system represents a frame-work aiming at the improvement of external coherence between individual programmes and between the supply-side and the demand-side of the inno-vation system. An interesting feature of the British study is the introduc-tion of new aspects of coherence, the relationship between the support system and the market.

Since the overall SMEPOL study has a regional as well as a SME per-spective, it is suggested that the concept of coherence needs to be expanded to cover not only a horizontal perspective – typically associated with the national level – but also, importantly, to focus on vertical co-ordination as well.

7.3.2 Regional Perspectives on Coherence

As indicated, the regional aspect of coherence is addressed differently in the national studies. The Danish study focuses on the importance of providing an adequate regional system of policy delivery. The pivotal role of Business Links in the UK system underscores the same argument. The regional application of national programmes in the Austrian study, as well as the Norwegian White Paper on regional policy from 1993, stresses the impor-tance of the regional focus of innovation policy, at least when looking at on-site implementation:

> The White Paper on regional policy, however, brings this into a wider framework based on an acceptance that knowledge is a key factor for the future industrial development, a tight coherence between the national strategies for enhancing competence and the regional political efforts. A key objective for the regional policy is to contribute to increased accessibility and improved exploitation of the national instruments. (Isaksen et al. 1999, chapter 1, p. 180)

The main objectives are to secure accessibility to national programmes for business in all regions, and not that all regions should develop a self-sufficient system of knowledge and innovation support. Correspondence between the policy programmes launched and the way in which SMEs understand these programmes, as well as correspondence between national R&D policies and regional innovation programmes, are the most important features of regional coherence. It is necessary to stress that we are dealing with demand-side as well as with supply-side coherence. For the evaluation of the specific programmes this means the programmes have to be seen in their specific context including the regional policy setting and the business community as well as the target enterprises of the programme. External coherence has to be viewed from a regional and a national perspective. The integration of related policy programmes may be provided at the national as well as at the regional level, but overall (regional) coherence can be achieved only if this condition is fulfilled. Qualitative differences may be conceived in terms of the regional innovation perspective.

7.4 COHERENCE RECONSIDERED

Uncertainty can no longer be considered an imperfection. It has to be seen as a basic condition for small as well as large firm innovation. Studies of successful regions suggest that actors and firms participate in loosely coupled systems of innovation, which co-evolve in dynamic ways. Firms' perspectives on their innovative activities are often highly interdependent. This is confirmed by the summary findings in Chapter 4. A few references will do:

- The Dutch study in Limburg thus showed a frequent problem for SMEs, especially regarding customer and supplier dependency. Firms mainly have a reactive and only rarely a proactive attitude towards input and output markets.
- The Norwegian study shows that there is a lack of co-operation with other local fish-processing firms and the regional R&D milieu. Technology-based firms have generally close co-operation with some large and demanding customers. The lack of close co-operation with local suppliers and other firms may hamper innovation activity.
- In the Belgian study of Wallonia, SMEs often envisage innovation from the point of view of improving processes or the use of new technology, but are not so well equipped to analyse market trends and opportunities.
- The Danish study in the Triangle Region shows that there is an implicit preference to form innovative relationships with customers as opposed to suppliers, even though firms feel that there is a lack of suitable customers to collaborate with. Internal barriers play a strategic role when it comes to the pattern of innovation – and the

innovation of new products is largely driven by the need to 'keep their customers'.

In such a dynamic context, a range of programmes may easily develop aiming at filling gaps and niches of envisaged needs. However, since a broad range of instruments may confuse the clients and lead to unintended slack in the programmes, it is seen to be of importance that the policy programming is flexible, i.e. is able to respond to changing contextual and situational conditions. Therefore, not only programming, but also the organizational framework, is of prime importance. This is seen to be in contrast to traditional ways of conceptualizing coherence, since the sole focus seems to be on programmatic coherence.

7.4.1   An Integrated View on the Coherence of Innovation Support Instruments

The concept of coherence that is reported in section 7.2 indicates that coherence of innovation policy is a heterogeneous concept which varies with administrative traditions and contextual matters. In practice the interactive innovation systems in particular seem to have a less straightforward internal (top-down) organizational set-up leading to lower scores for this dimension. Therefore learning aspects in the set-up of innovation support schemes as well in the co-ordination of programmes and delivery have gained importance.

The national studies point to a number of aspects of importance when it comes to the governance, guidance and organizational implementation of successful innovation support programmes in a dynamic setting. A few of these experiences are summarized below, with the aim of highlighting different aspects of the policy processes and system building:

1.   A number of programmes demonstrate that overall support paradigms followed have evolved out of political and institutional traditions in the country, which were highly context-specific and idiosyncratic. The paradigms may have guiding values at different levels of the programme organization. The Danish case of the technological service system may illuminate the role of common paradigms for coherence of activities. In Denmark the Approved Technological Service Institutes (GTS) have evolved into a dual system based on a long tradition of public support for the diffusion of new technological knowledge. The peculiar thing about the system is that it has, for many years, had to achieve a balance between a public utility function on the one hand, and a public demand that the institutes' activities should be run on

commercial – non-distorting – lines, on the other hand. As Christensen (1996) writes:

> 'It is not at all obvious that a country should have such institutes. Actually it is only a few countries, which have a similar structure with a network of institutes, which are neither purely market-based nor massively funded by the state.' (Christensen 1999, 10)

2. Another issue is how support paradigms are followed and their guiding values at different levels of the programme organization. A good example of the alteration of principles, according to the tier of government, is Austria. Instruments offering direct support such as grants and loans dominate the national innovation support system of Austria. The funding programmes and the responsible institutions are organized along the linear innovation model, providing financial resources for certain stages in the innovation process (research–development–commercialization). At the regional level, innovation and technology policy focuses more on elements of the interactive innovation model. At present, this applies primarily to technology centres. Most centres concentrate on incubation, but some also comprise R&D and training/education. More recently, technology transfer and the stimulation of clusters have become new important strategies of the regional government.

3. The next issue is how the learning perspective is included in the system and brought into operation. It can be difficult to highlight the learning perspective implied in the processes framed by innovation support schemes. However, the Norwegian case presented in section 7.3.1 is an illustrating case in point. In this case internal learning processes among actors involved in the programmes tended to lower external coherence of the instruments implied. Another case in point is found in the UK country report in the British SMART Award Scheme. The programme provides a good example of policy learning as a result of experiences gained from implementation since the scheme was first piloted in 1986. Policy learning occurs both as a result of formal assessment and of more informal and ongoing interaction between government officials responsible for the Scheme within each Regional Government Office and with recipient firms. Inter-regional policy learning is also furthered through interaction and sharing of experience between government officials in different regions. There is little evidence, however, of policy learning based on interaction with advisors and organizations outside the SMART Scheme itself.

    There have been a number of recent developments aimed at addressing some of the identified shortcomings of SMART. The most important is that SMART winners help to address some of the problems commonly experienced by innovative small firms facilitating mutual support and learning between award-winners. Following earlier criticisms of the overly stringent conditions of the Scheme, in 1999 the SMART scheme was expanded to include three further elements: (i) Micro Projects – low cost technology-based innovations in businesses with fewer than 10 people; (ii) Technology Reviews – expert reviews to help SMEs benchmark themselves against best practice technology in the sector; and (iii) Technology Studies – for SMEs to help identify technological opportunities which may lead to innovative products and processes. This is likely to have the effect of making the Scheme more appropriate to a wider range of SMEs, and to reinforce learning cycles.

4. How innovative dynamics are envisaged is highlighted in Norway. At a general level, the programmes aim to generate new behaviour and knowledge in firms (and also in R&D institutions and colleges in the cases of TEFT and RUSH). That is, the programmes aim to teach firms to use R&D institutions and to organize and carry out innovation projects. The programmes' methods and activities reflect these targets; attachés and case handlers visit firms, analyse firms' innovation needs, recruit firms to the programmes. Finally they connect firms with relevant R&D milieus and researchers and support firms' innovation projects.

5. The changing role of vertical co-ordination is highlighted in a number of country studies. Vertical co-ordination, defined as the links between bodies at the national and at the regional level, can be obtained in different settings. Vertical co-ordination in Spain within the region (Comunidad Valenciana) between the IMPIVA (regional public policy fostering industrial development in the region), the Technological Institutes (regional level), and the firms is an obvious example. Moreover they are also involved in national and EU policies through the channelling and management of national and EU funds in order to finance innovation activities within SMEs in the region.

   With regard to the Technological Institutes in the Region of Valencia an example of vertical co-ordination within a region is the collaboration between the IMPIVA organization and the Technological Institutes, as well as the links between firms and the Technological Institutes, which are strong advantages of these policy tools.

6. The changing role attached to vertical co-ordination is illuminated in the case of the 15 regional Technological Information Centres (TIC) in Denmark. The Danish Technological Institutes (DTI) have run them since 1974. In 1996 they gained a separate legal and operational status. In order to secure a uniform nationwide service and in order to support the development of new information support programmes, the 'TIC-

Denmark' was formed in 1996 as a co-ordinating body. However, in 1999 it came to an agreement with the government and the union of county councils that the county councils were to fund half of the budgets provided for each of the TICs.

In exchange, the county councils gained decisive influence on the activity profile of the centre located in their region. In conclusion, a highly diversified regional support system is evolving at the expense of a uniform system based on national defined objectives and activities. A regional integration of innovation schemes aiming directly at SMEs in the region tends to substitute vertically co-ordinated national schemes. In consequence the 'learning cycle' co-ordinated by 'TIC-Denmark' is vanishing, while a tighter learning regime between actors at the regional level is expected to evolve.

7.   Various modes of regional co-ordination have been shown throughout the studies. A prominent feature in the Norwegian innovation support system is 'TEFT'. This body has a special role in the support system at a regional level by co-ordinating its instruments with other policies and programmes. This is supposed to occur when the attachés visit firms and analyse their situation and needs. Firms should be directed to policy tools and to R&D institutions other than the four participating in the programme. The NT programme also relates to other initiatives. TEFT in Northern Norway is integrated into NT, and NT co-operates closely with FORNY (a national programme of commercialization of research results in R&D institutions) as well as with regional programmes like MABIT (a research and industrial development programme in marine biotechnology). Recent reforms in the Danish regions have highlighted regional co-ordination between the traditional state-operated TICs and the regional business development agencies.

The way innovation support programmes are organized differs from country to country as indicated in the above examples. The degree of central governance and funding is usually closely related to the general style of government, see Figure 7.1 above. Sectorial programme co-ordination can be found in all countries, and is often found to be in contradiction to regional needs, as can be seen in some of the Danish and Austrian examples mentioned above, but also exhibiting degrees of variation from region to region as well as within countries.

7.4.2   Theoretical Perspectives Inferred

Inspired by Mintzberg and Waters (1985), who emphasize the evolutionary aspects of strategy-making, policy programming can be envisaged as a pattern, i.e. a stream of actions. By this definition innovation policy programming is seen as 'consistency in behaviour, whether intended or not' (Mintzberg et al. 1995, 14). The emphasis on consistency in behaviour gains importance when talking about a policy-producing system which comprises several autonomous organizations and where a dynamic perspective is employed.

Also, the implication of Mintzberg and Water's notion of 'a stream of action' is that policy programming cannot exclusively be seen and understood as a plan, or an intended outcome, because unintended action will always appear and interfere with intended plans. Policy programmes and their realization are thus the combined result of intended (or designed) activities and emergent actions. Therefore, we may envisage an intended policy programme, which is designed with coherence strongly in mind, and a realized policy programme, which consists of intended as well as unintended elements, i.e. elements that emerge in spite of, or in the absence of, intentions.

Basically, it is hard to think of a policy programme being realized in full accordance with intentions realized. It is also hard to think of a policy programme where the outcome is solely based on random inputs, i.e. where intentions are fully absent. Therefore, in practice there will always be a larger or minor gap between intended (internal) coherence and coherence ex post. Granted the proposition that coherence is a key to efficiency in policy programming, those factors producing the gap between coherence ex ante and ex post are of crucial importance in a dynamic setting. Therefore, any evaluation attempt to measure programme efficiency thus means that efficiency has to be measured against ex post and not ex ante coherence according to the concept of this chapter. Bounded rationality based on actors' diverse positions in policy programming is probably one of the most critical factors in systems with many actors and contexts. Their view on what ought to be the intended path will, inevitably, tend to differ.

However, seen as a stream of action, two pathways tend to prevail in policy programming. In one path the issue of responsiveness is central. Huge differences in the regional industrial dynamics as well as administrative and political traditions founded on a philosophy of regional autonomy may promote this strand. Also a stronger element of socio-economic change may favour this strand. In the other stream, central co-ordination is a key issue. It may be caused by traditions favouring a strong central planning agency or by the squeeze on public spending and claims on public service provision.

In Figure 7.3 below the 'idealized' path of intended support policies designed is illustrated as an intended stream of action in which strong co-ordination goes hand in hand with a strong element of responsiveness

(Path I). From an ex ante perspective, coherence is then very much a question of how to integrate streams of action with diverse actors centrally as well as on a decentralized basis. From a combined perspective, coherence in policy programming is then how to frame and guide behaviour in such a way as to balance responsiveness to SME clients' needs with administrative needs of co-ordination. In some national innovation systems a strong tradition for responsiveness prevails (Path II). In other nations, central co-ordination is seen as the guiding line (Path III).

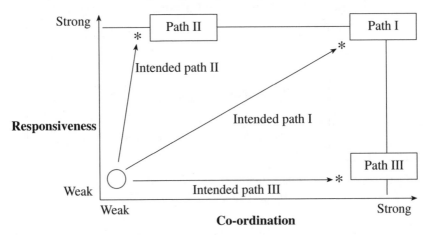

Figure 7.3   Different intended paths of support

However, although all nations ex ante may basically work out similar intended paths of programming, the stream of action is influenced by a number of forces embedded in the context of each nation. Therefore, realized programmes will deviate from intended programmes in a number of respects, as visualized in Figure 7.4.

As already indicated in Chapter 3, there are pronounced differences in the way innovation support policy programmes are organized in the nations participating in this study. Differences in the administrative traditions and differences in the organization of territorial space are major causes of these differences. At the same time differences in, for example, the balance of power between the state – including different national bodies and agencies – on one side, and local and regional bodies involved in regional industrial policy issues that focus on SMEs on the other side influence the way support programmes are realized.

Some nations have a long tradition and a strong political agenda for a high degree of (intended) regional responsiveness (Norway), although programmes are nationally initiated, designed and executed. Other countries seemingly have a weak tradition for regional initiatives (UK, Denmark), although policy programmes included in the UK case do demonstrate how national and even international initiatives may be realized in response to regional initiatives and needs.

In yet other countries the regional bodies seem to be so strong that intended support programming is biased towards regional responsiveness at the outset. In such cases the issue of co-ordination tends to be realized in a regional more than a national context. This is for example the case of the Spanish programmes on Technological Institutes in the Valencia region.

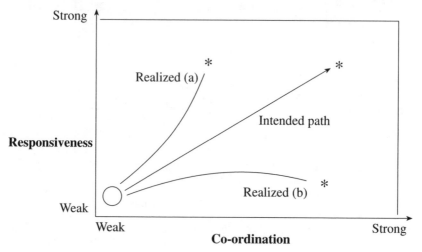

Figure 7.4   Intended and emerging support programmes

The Austrian innovation support schemes realized are positioned along different lines in Figure 7.4. In accordance with the nature of most of the programmes being national ones applied in a regional context (top-down) they tend to be realized with a strong emphasis on co-ordinated action. The TCs alone are realized almost along the intended path (centre) with inter-mediate levels of responsiveness and co-ordination.

Regarding the Danish innovation programmes, the TIC is of special interest, since a major change in intended path of support programming has taken place. The TIC has moved from the lower right (b) to the upper left (a), reflecting the fact that bottom-up features are becoming more and more predominant following reduced state involvement in TICs. The question is if this change in intended path will also lead to greater responsiveness to SME needs, or only lead to growing responsiveness to the regional policy agenda, including those financial restrictions following from a regionalization.

In the Walloon programmes, neither responsiveness to changing SME needs nor co-ordination with other programmes are judged to be of major concern. In this way the Belgium study demonstrates a case in which an intended path tends to be realized almost as expected, due to a restrictive programme not open to adaptation. But they are all marked to be realized in the lower left corner of Figure 7.4, indicating that responsiveness as well as co-ordination is weak.

The Dutch study demonstrates a case (the so-called KIM scheme) in which co-ordination as well responsiveness to SME needs have been growing. It also included a scheme (KIC) in which a growing responsiveness to large firms' needs was realized, while other elements of the scheme were integrated with other support schemes targeting SMEs.

In contrast to the Belgian schemes in which the regional organization of the programmes has no role to play at all, the regional programme organization is seen to play an important role for a successful implementation of schemes in the Dutch case. In the Italian as well as in the Austrian case, the role of the regional programme organization is judged to be much more mixed for a successful implementation of programmes.

In the realized stream of action favouring responsiveness, external coherence tends to be considered a key issue, while internal coherence tends to be considered of prime importance in systems where the regional programme organization is seen to have a modest role for a successful implementation.

7.5   CONCLUSIONS

The concept of coherence is only weakly represented in public planning theory. Likewise, the concept is largely absent in the theory on strategic planning and management. It is, seemingly, mostly used in applied evaluation studies and it seems to be strongly linked to traditional conceptions of policy programming. The national studies suggest that the concept of coherence is useful in application, but that it has to be conceptualized in a dynamic as well as a spatial setting.

The national studies most often stress huge differences between intentions emphasized in the programmes and the stream of action realized. Therefore it is a major conclusion that organizational coherence is of greater importance than ex ante programmatic coherence. Several reasons for this can be found.

First of all, it was found that organizational coherence is an essential prerequisite for the creation of intensive communications about the intentions of policy-makers and the views of regional bodies and actors on regional needs and implementation policies in respect of the delivery of programme packages.

Secondly, the organizational set-up constitutes an important element of continuity in a dynamic policy context, where innovation policy programmes tend to change very often. Phrased in other words, the more responsive the innovation policy programmes are in their design to meet changing needs and foster dynamic processes and interaction at the regional scenes, the more importance must be attached to the organizational coherence of the system carrying the programmes. However, organizational coherence should not be confused with organizational rigidities.

Thirdly, and related to the issue of continuity, the creation of organizational coherence is important for stimulating learning among actors, programming funding, and motivating actors involved in the implementation of programmes.

Fourthly, in a dynamic setting there is a need for programmatic flexibility in order to respond to changing or diversified needs of client enterprises. This flexibility can in principle be by way of an organizational set-up that favours programmatic integration of key support tools that target different barriers to innovation among SMEs at the regional level. In terms of the 'stream of action' perspective taken above, flexibility may thus be attached to two separate phenomena, namely, the flexibility in programming (i.e. the acceptance of deviation from intended pathways) and the organizational flexibility.

The extent to which clients' views, and the views of those involved in up-front delivery system, are included in the programming, is of importance, since these views act as guideposts for programmatic performance in several respects, not least that of learning alongside the chain constituting the innovation support delivery system, including the clients.

PART IV

SMEs, INNOVATION AND REGIONS:
DESIGNING POLICIES

8. Towards a new paradigm for innovation policy?[9]

Claire Nauwelaers and René Wintjes

8.1 INTRODUCTION

Innovation ranks higher on policy agendas today, at national, European and regional levels. This evolution is nurtured by the understanding that innovation is the key to economic development for advanced, high-wages countries. It is becoming visible through a gradual shift in policy statements from support for R&D and technology diffusion to the promotion of innovation. The understanding of innovation as something different from R&D and the diffusion of technology is gaining ground: innovation refers to the behaviour of enterprises, planning and implementing changes in their practices in order to come up with new products, processes, services or organization. This change in focus reflects efforts based on the view that innovation is an interactive, rather than linear, process (Rothwell 1992). On this understanding, traditional science and technology policies do not offer the unique response needed to support innovative practices. Instead, many other elements in addition to science and technology play a role in innovation and need to be tackled by innovation policy (Soete and Arundel 1993; Cowan and van de Paal 2000). Envisaged in such an enlarged framework, innovation policies are still in their infancy.

The thesis at the core of this book is that the variety of regional contexts, the diversity in firms' abilities and attitudes, and in driving forces and barriers towards innovation, prevents the search for one permanent 'best practice' policy, valid for each and every situation. This is not to say, however, that nothing general can be concluded in response to the question of how to improve the efficiency of policy instruments to support innovation in SMEs. Rather, the results from the analyses of a variety of innovation policy tools, based on the same conceptual background, provide an answer to this crucial question. The various tools form a rich scope of opportunities for better practice regarding the policy process of addressing innovation of SMEs in their regional context. If one may call the shift from a linear model of innovation towards an interactive one a shift in paradigm, then the main goal of SMEPOL is to provide evidence for a similar shift towards a new innovation policy paradigm. The aim of this chapter is to bring to light the main elements of such a new policy paradigm.

The points of departure of this policy-oriented study are that innovation is a good thing (both on the regional and firm levels), and that there is a call for public intervention in order to get more of it. As a background, based on the literature, the first two chapters build on these basic assumptions, which include two main arguments.

First, Chapter 1 argues that SMEs are an important target group for innovation policy. It provides three distinctive characteristics of SMEs, which form the basis for arguments and implications for innovation policy directed to SMEs. These distinctive characteristics (compared to larger firms) are: a limited resource base, a distinctive organizational culture linked to the proximity between ownership and management, and a lower ability to shape their external environment. These characteristics, which call for distinctive policy approaches, are at the root of the more informal, uncodified character of management and innovation practices in SMEs. Also, this chapter warns against bold generalizations on 'the' SME: technology-driven, technology-following, and technology-indifferent SMEs have very different needs and aptitudes, and this calls for different types of policy intervention. According to this view, the thrust of policy approaches should be twofold: to increase the availability of external resources for SMEs and to develop their internal absorptive and learning capacities (Cohen and Levinthal 1990). This points to: (i) the crucial role of intermediaries who, working on the basis of personal trust relations, are able to codify SMEs' needs; (ii) the value of 'peer' networks as learning channels; and (iii) the key role of human capital in SMEs.

Second, chapter 2 asserts the importance of the regional dimension of innovation. The discussion develops the thesis that proximity linkages can be instrumental in developing 'learning firms' and 'learning regions'. Also, broadening and extending the concept of clusters towards one of coalition development points to a broader scope for innovation policy, that of supporting the social and cultural aspects of innovation and enhancing social capital as a key element behind well-functioning regional innovation systems. Developing collective capacities and networking practices at the local level follows logically from this policy aim, but this goal should also be complemented with support to the development of linkages at national and international levels, in order to avoid being trapped by too strong local ties, possibly leading to lock-in situations.

For the translation of broad policy orientations into operational guidelines, we use the findings of the SMEPOL empirical analyses, as developed in the subsequent chapters: Chapter 3 for the analysis of the variety of regional contexts, which we use to build up context-sensitive policy recommendations, Chapter 4 on the patterns of innovation in SMEs, which illustrates how the generic attributes of SMEs translate into specific barriers and assets for innovation, and Chapter 5 on the typology of policy instruments, which allows us to develop the main thrust of our policy argument into more precise guidelines for each type of instrument. To build up our recommendations, we also consider the reflections on results and impacts, and 'good practice elements' of policy tools, as developed in Chapter 6, as well as the thesis put forth in Chapter 7 on the importance of combining responsiveness and co-ordination in the programming, organization and implementation of policy.

The present chapter is organized as follows. Based on the findings achieved in this comparative research, sections 8.2 and 8.3 lay the claim for a new innovation policy paradigm and propose a shift in rationale (section 8.2) and in the broad orientations (section 8.3) of innovation policy to addressing SMEs in their regional context. These arguments are based on the notions of market failures and system deficits, which show up as barriers to innovation processes in regions and SMEs, and on what we may call government failures, shown by the evaluation of instruments. Using these arguments, section 8.4 discusses how the SMEPOL instruments could be improved, drawing lessons from good practice. The key message of this chapter is synthesized in section 8.5, where a stylized view on the content of a sound regional innovation policy for SMEs is presented. Section 8.6 deals with the question of how to build a coherent portfolio of policy instruments, taking into account both regional situations and specific SMEs' needs in terms of innovation. The message delivered is that there is no 'one-size-fits-all' policy portfolio. The concluding section draws the lessons from the whole exercise of evaluating, in a comparative fashion, a variety of policy tools, within a common conceptual framework. One salient element of the conclusion is the need for more 'policy intelligence' in this complex field.

8.2 THE RATIONALE FOR POLICY INTERVENTION: FAILURES, BARRIERS, AND BOUNDARIES

Whether we talk about markets, systems or governments in relation to innovation, it all concerns communication, a process of exchanging information and knowledge. In order to be useful and valuable to others in a firm, market, system or government administration, technological (and other) knowledge has to be diffused and policy lessons have to be learned.

The typical and traditional approach to communication in economics focuses on markets where price mediates supply and demand whereas, the 'neo-classical' government typically communicates power based on a hierarchical position vis-à-vis the economic agents they govern. In accordance with traditional market-hierarchy dichotomies, the typical argument for government intervention is when markets fail in communication. Either the market or the government would provide the best solution. In general, interactive communication is not considered to be of vital importance in the process of finding and reaching solutions. This linear perspective prevailed even before it had been applied to innovation.

When everybody knows in advance what (products, resources, technologies, capabilities etc.) are being talked about and everybody would agree on its (present and future) economic value, the market is perfectly able to communicate supply and demand. In these situations there is no need for interactive communication. Whether it is the 'demander' or the 'supplier' who names the price, the market will, in a linear response, come up with the proper answer. For the exchange of certain goods or services, the price may be the only aspect that has to be communicated. However, where knowledge or innovation is concerned, the price mechanism may not function very well.

Following the same logic for policy decision-making, a central question for a policy-maker is: how do I recognize where and when markets fail, so that I know where and when to intervene? If it is perfectly clear to policy-makers where markets fail, and it is widely agreed upon what the governed region additionally 'needs' and 'has to offer' (more specifically, what firms 'need' from their region, including government, and what they have to offer their region and its policy goals), then there is no need for interaction. Everything is clear: there is no knowledge left to be codified, there is only information to be passed on. Since interaction is costly in terms of time and energy, linear and top-down communication is likely to be more efficient.

However, to deal with the uncertainties attached to knowledge and innovation (Dosi 1988), economic and policy agents may want to communicate more than price or authority. The traditional concept of markets and (state) hierarchies with their anonymous, linear and formal communication, fails to incorporate these broader information needs. A reason why markets as co-ordination and communication mechanisms may not function very well regarding innovation is related to the uncertainties attached to predicting the future (Cowan and Foray 1997). The market may fail to predict the economic value of new technologies, new products, new resources, new firms or new entrepreneurial capabilities. Typically, the market will, for instance, not be able to value a start-up firm. Although policy-makers also have difficulties in predicting the future, this kind of market failure is a widely accepted justification for public intervention. For example, a generic national policy tool, like a tax-reduction scheme, seems relevant to 'protect' these young, new entrepreneurial experiments, providing them a chance to prove themselves and to convince the market (that is, customers, but also financial and labour markets) of their potential and, moreover, to convince the government of their potential contribution to the region and its policy goals. The same arguments may hold for new sectors or technologies or a young regional cluster of firms, or even older non-innovative firms that want to and are trying to become innovative.

Other places where the market obviously fails is in communicating certain environmental and social costs and benefits. If economic agents do not take these kinds of 'costs' into account, governments may want to intervene and extend the boundaries of the rationality of the agents they govern, e.g. by influencing their cost–benefit calculations with environmental taxes.

The justification for traditional technology or R&D policy has been put forth by Arrow, and is based on the macro-level argument that when social effects are taken into account, there is under-investment in R&D. The risk and uncertainty attached to R&D by private actors calls for public intervention because, at the macro-level, it is considered worthwhile to publicly take the risk for the sake of society, e.g. by financing public R&D in universities or, again, by influencing private, micro-level cost–benefit calculations with tax deductions or subsidized facilities.

The idea that there is a role for policy-makers, if markets fail, does not imply that policy-makers are perfect, but that the above-mentioned general or structural market failures may very well be effectively and efficiently addressed by generic policy instruments, designed and delivered at the national policy level. Further, to diffuse information on needs and support, linear communication seems appropriate. However, knowledge differs from information. For instance, distance does not seem to be a barrier to the transmission of information, but in the transmission of knowledge it does.

The importance of the tacit dimension (Polanyi 1996), the informal, uncodified and disembodied aspects of the knowledge concerned, both at the regional level and for SMEs, underscores the localized nature of knowledge spillovers (Storper 1997). The linear communication arguments using the old market–hierarchy approaches fail to address this. Proximity matters to knowledge spillovers and interaction between regional agents (both private and public) matters in dealing with the uncertainties attached to innovation processes in regions and SMEs. The variety of situations regarding innovation, SMEs and regions call for communicative interaction (Nauwelaers and Morgan 1999). Local discussions, private and public–private ones, can shed more light on the uncertainty issues. Exchanging tacit visions, converging ideas and co-ordinating investment decisions (public, private and public–private ones) may provide the knowledge base for an innovation strategy concerning SMEs and their regions (Lundvall 1996).

Especially concerning innovation processes in regions and SMEs, the concept of systems (or networks or clusters) seems a more realistic model to follow than the traditional concepts of markets and hierarchies (Grabher 1993a). A regional innovation systems approach stresses the importance of the diffusion of knowledge and interactive learning within the region as a system (Morgan 1997). The non-anonymous relations, the complementarity of activities, and the historical setting are stressed in specifying the regional context and the profile of the region's SMEs. These give the regional system its identity, so to speak (Feldman 1994). Further, in order to find out and articulate what a particular region or firm needs, or what is lacking concerning innovation, regional proximity and communicative interaction are appropriate to address the tacit and latent aspects of such needs (Landabaso 1997). Providing R&D tax reduction or subsidies may not be enough to change the rationality (nor its boundaries) of SMEs regarding innovation processes. Those arguments underpin the growing importance taken by cluster policies, as the concept seems to fit quite well the prescriptions of an interactive view on innovation (Nauwelaers 2001).

8.3   LESSONS FROM EVALUATING 40 INNOVATION POLICY TOOLS IN 11 EUROPEAN REGIONS

Having discussed the rationale for policy intervention in innovation, this section deals with the content of such policies, proposing general principles for their design and implementation (section 8.3.1) and observing how these principles compare with practice in the case-study regions (section 8.3.2).

8.3.1   Basic Principles for Innovation Policies

Building on the conceptual and empirical findings of the SMEPOL project, this analysis puts forth the following proposition: since the main distinguishing features of the majority of SMEs, with regard to the innovation process, are that they have a limited resource base, they need external orientation to understand and (pro-actively) adapt to their environment, and they engage in innovation in an informal mode, therefore the main role for innovation policy, which aims to increase the capacity of a region and the capabilities of its SMEs to innovate, is to foster interactive learning within the firms and within the region. This calls for an interactive mode of policy intervention.

Of course, this statement stands as quite a bold generalization of both SMEs' characteristics and policy challenge. Some SMEs have quite an advanced knowledge base (e.g. new technology-based firms), others have developed excellent innovation management capabilities and explicit innovation strategies, and some firms in niches really shape their business environment rather than being dependent on it. Also, there are problems of another nature which keep SMEs from innovating, such as the need for risk financing[10] or the necessity to access state-of-the-art technology. However, the meaning of this proposal is to point to new orientations of policies to address the key needs of the majority of SMEs in most regions, which are not properly taken into account in traditional policy approaches. This does not imply that linear approaches and tools are not relevant any more, but rather, it puts the latter in perspective. It means that providing resources to innovate (finance, technology) is not sufficient if the firms do not possess the managerial and organizational abilities to deal with the innovative process (Cobbenhagen 1999). The view of an 'automatic' flow of technological resources through the firm, or from the R&D sector into the firm, is argued against here, while increased attention is given to the innovation process (within and around the firm) itself, in a broader sense (Nauwelaers 2000).

Nevertheless, fostering interactive learning, as a policy goal, should not be read in a dogmatic egalitarian sense, limiting the view to the development of 'Third-Italy' type of horizontal networking and relationships as an ideal way to foster that process. Hierarchical relationships might be very relevant ways to achieve such an objective too, depending on the environment. As developed in section 8.2, the role of geographical proximity might be important to nurture learning relationships, but it is not a necessary ingredient everywhere. The point here is that being open to outside sources of knowledge, and having the capacity to integrate these with internal knowledge in the firm in a continuous mode, is key to the innovation process. Such an objective has implications both on the supply side (outside resources should exist, be organized and accessible to firms) and on the demand side (the firm's absorption capacity and its willingness to entertain links with the outside should be enhanced). Developing strategic capabilities, at the firm, organization and policy levels, lies at the heart of this challenge.

The idea of an interactive mode of policy implementation means not only that services should be both designed and delivered in co-operation with the beneficiaries, but also that the policy implementers can be partners in the supported action or project, so that learning can happen both ways – between policy implementers and firms, that is what we called 'communicative interaction' in section 8.2 above. In this approach, the tacit nature of innovation in SMEs is better addressed than in more hierarchical policy modes.

8.3.2 Application of the Basic Principles

With the above proposal as a theoretical challenge for policy, how does it compare with practice, as experienced in the regions covered by the SMEPOL study? The horizontal overview and comparison of the analyses of 40 policy tools in 11 European regions shows that such a challenge is hardly met by the actual policies. Both the content and the modes of delivery of policies are in most cases not interactive and fit better with a market-hierarchy than a system approach. More precisely, the SMEPOL analyses deliver the following picture:

1. The general situation is that linear tools are dominating the policy scene, but that everywhere an evolution towards more interactive support is visible. The following quotes from the national studies illustrate this point.

   > In Belgium: 'One can analyse the actual state of the emerging Walloon innovation policy as follows: this policy is in fact founded on two different paradigms – broadly-speaking the linear and the interactive views on innovation – the former being embodied in the "mainstream" policy while the latter is translated in the "fringe", an unarticulated and rather fuzzy set of initiatives, trial-and-errors efforts, inspired by the "localized externalities" approach and much less linear in scope' (Nauwelaers et al. 1999).
   >
   > In Austria: 'The Austrian innovation support system is dominated by a few funding organizations, mainly offering direct support like grants and loans within the framework of several programmes. Institutions and programmes are organized along the linear innovation model. [However] There are serious doubts about the efficiency of traditional direct support for R&D and innovation' (Kaufmann and Tödtling 1999).
   >
   > However in the Netherlands: 'The "interactive" content of Limburg's innovation policy is largely due to the RTP Limburg initiative. The RTP framework and the way Limburg has implemented it has led to an extension of the "interactive" policy. "Interactive" instruments are not "delivered" on paper at the front door, but are mainly implemented in personal communicative interaction with the actors involved. The regional intermediates Syntens and LIOF play a vital role in the implementation of Limburg's "interactive" policy' (Nauwelaers et al. 1999).

2. A set of policy instruments, in general, does not form a system: lack of co-ordination and of synergies among tools, or 'lack of external coherence' in the words of Chapter 7, is the rule.

   > In Italy: 'It seems that there is a lot of overlapping in missions of different institutions working within the Apulia region, with a lack of capability of co-ordination for what concerns the Apulia Region institution' (Garofoli 1999a).

In Denmark: 'None of the actors have made strategy- and action-oriented links between the forthcoming knowledge-based economy and the attempt to formulate a coherent learning and innovation programme. Most initiatives are of a single programme nature. A stimulating strategy trying to integrate and co-ordinate the diverse innovation schemes is lacking' (Christensen et al. 1999).

3. Few policy instruments are designed and implemented in an user-oriented mode, taking both expressed and latent needs of users into account: the majority of tools are developed in a reactive, top-down fashion and at best consider firms' needs expressed (but not latent). However, in cases where 'voice' of users (i.e. firms' expressed, or even latent, needs) is taken into account, the tools become much more user-oriented.

In Italy: 'Tecnopolis represents a typical model of technological park based on supply side, starting from the existence (and redundancy) of competences (. . .), always postponing the moment of monitoring the potential needs of local (or external) firms to facilitate interaction between research centre and economic activities' (Garofoli 1999).

However in Spain: 'Local entrepreneurs belong to the "Consejo Rector", the governing body of the Technological Institutes, so that they participate in the design of activities and policies (. . .) The Institutes' strong points are their nearness to firms, their connections to other international centers of this type and the knowledge transferred to the firms through them (. . .) The Technological Institutes show a high degree of effectiveness in adapting to innovation support needs as expressed by the entrepreneurs of the SMEs in the four sectors studied' (Vázquez-Barquero et al. 1999).

4. Policy learning is still rare and underdeveloped (Nauwelaers 1997). If it occurs at the level of organizations, it takes place in an occasional, not routinized way. Intense policy-learning practices may, however, result in undesirable volatility in the policy system. At the other extreme, it seems unjustified to maintain a range of tools that are virtually not used by firms. The challenge lies in fine-tuning the policy tools without letting firms suffer from the instability of the system. This trade-off between responsiveness and organizational coherence has been discussed at length in Chapter 7.

In Austria: 'Most support instruments are not evaluated systematically. Of course, there is an ongoing learning process in the institutions about the effects of support activities, about problems and needs of firms regarding their innovation processes. But it is based on personal experience and information exchange, the learning process is not institutionalized or routinely organized' (Kaufmann and Tödtling 1999).

However in the Netherlands: 'National policy-makers have learned from several regional innovation policy initiatives and integrated them into the national innovation policy. The analysed KIM – and KIC – schemes are good examples from Limburg of this bottom-up policy learning process' (Nauwelaers et al. 1999).

However in Norway: 'Policy learning takes place from evaluations in the support system. Thus, TEFT and NT have been changed during their "life", partly as a response to knowledge acquired through evaluations. TEFT also carried out monitoring research, which is implemented in the new REGINN programme too. RUSH, being an experimental programme, in particular, would need procedures for systematic evaluations and learning, however lacking in this case' (Isaksen et al. 1999 chapter 4).

5. There is an emerging new tendency of developing 'overall schemes', gathering into a single programme, instrument, or organization a set of tools that traditionally are proposed separately to companies. This approach is promising in that it fits well with the global perception of innovation within firms: it impinges on all activities of the firm.

In the Netherlands: 'The 18 Syntens organizations are the most important intermediary organizations in the Netherlands (and thus Limburg) for innovation policy addressed at SMEs. The role of the ICs gradually changed from bringing technology to regional SMEs (technology push) to an intermediate role of "broker", and more recently it fulfils the role of "organizer, animator or coach". The "new" organizations are more able to deliver "all-round" service and support to SMEs and they go by the new name of "Syntens: an innovation network for entrepreneurs"' (Nauwelaers et al. 1999).

Example from Norway: 'NT provides also a more all-round support for innovation than the other programmes that concentrate on a single component in the innovation system . . . NT focuses on firm's innovation projects, and tailor-made support to firms' specific needs, both technological and non-technological support' (Isaksen et al. 1999 chapter 4)

6. The majority of instruments aim at improving or facilitating existing innovation projects, rather than inducing new innovation practices. Providing grants for R&D, for example, seems to induce a rather small incremental behavioural effect (in terms of changing strategy, management or culture regarding innovation, co-operation and inter-active learning). Therefore, the value-added of such policy instruments, seen from a regional perspective, is questionable. It follows from this that the question of penetration rate of the tools in the business sector is not always addressed in policy settings. For example, where a 'picking-the-winner' approach is taken, a focus on the visibility of results may act to the detriment of the value-added of the scheme.

In the UK: 'There is a mismatch between the characteristics of entrepreneurs and SMEs in the Lee Valley area . . . and the eligibility criteria for the (SMART) scheme. Its competitive nature, using national assessment criteria, which are weighted towards new to the industry innovations, has meant that relatively few projects from within the Lee Valley have qualified. As a result, the SMART scheme is not making much contribution to raising the level of innovation in SMEs in this area' (Smallbone et al. 1999).

In Austria: 'In general, financial support seems to be actually "overeffective" (. . .) This is especially obvious in the case of the more innovative SMEs. This raises the question if there is a significant share of the applied funds lost to firms and projects, which do not really need them, but take them along as a welcome additional source of funding. A special problem of the direct support programmes seems to be the emergence of long-term stable relations to a special clientele consisting of well-known innovative firms' (Kaufmann and Tödtling 1999).

7.   Very often, tools designed at regional level work under a closed vision of the relevant sources of knowledge useful for firms, as the boundaries of the system are defined in administrative terms. If tools would be more user-oriented, there is no need for such a restricted view.

In Belgium: 'Most of the "fringe" initiatives work under irrelevant geographical limits, of an administrative nature, and imposed by the sources of public funding: provincial limits, Objective 1 or 2 zones, (. . .) which do not necessarily correspond with the natural areas of actions of the targeted firms' (Nauwelaers et al. 1999).

In Denmark: 'The national/international client issue tends to overshadow the issue of a regionalized technological service system. Most respondents approached tend to disregard the regional issue as a relevant issue with the argument that the service units will be too small. If every region is going to have a service centre, then the GTS system will lose economies of scale as well as of scope' (Christensen et al. 1999).

8.   Overall, there is an uneven degree of policy concern, among the five following innovation-support needs reported by SMEs:

●   Finance/risk sharing
●   Technology/technical know-how
●   Qualifications/personnel
●   Market access/information
●   Time constraints/Organization/Strategic capabilities.

The lack of 'market orientation' of the policy tools, or their lack of focus on the commercialization aspects of innovation, are particularly evident.

In Belgium: 'There is a need to reinforce further the human skills component of innovation. The take-up of existing human resources schemes is quantitatively limited, so there is a need to also upgrade externalisation capacities of firms to other sources than universities or knowledge source institutions, to favour better commercial capacities within the firm, and to support strategic capacities for SME managers on a wider scale, introducing more innovation management tools and creative thinking in SMEs (Nauwelaers et al. 1999).

However in the UK: 'The emphasis of BICs is on the commercialisation of innovative ideas. In this respect, BICs aim to provide a comprehensive package of support for innovative new ventures and existing projects. In emphasising the commercial application of innovation, BICs also aim to address the weakness which has been consistently identified in small technology based firms, of an overemphasis on technical development at the expense of marketing and general management skills' (Smallbone et al. 1999).

8.4.   STRENGTHS, WEAKNESSES AND WAYS TO IMPROVE INNOVATION POLICY INSTRUMENTS

The preceding section drew horizontal conclusions from the analyses carried out on innovation policy tools in 11 European regions. Broadly speaking, the message was that the general principles for sound innovation policies were not met in most of the cases studied. In this section, we try to go further by asking the following question: how could the tools studied, taken individually, be altered so as to get closer to these principles?

In Chapter 6, the innovation policy instruments have been classified, according to their nature, under five types, reflecting the different goals and targets of these policy instruments. We use this typology for our reflection:

1.   Direct support schemes for R&D and innovation projects;
2.   Technical personnel introduction schemes;
3.   Technology centres and schemes fostering technological diffusion to SMEs;
4.   Mobility schemes for researchers;
5.   Innovation Brokers and innovation advisers.

For each category, we list the relevant schemes studied in SMEPOL, we propose a summarized view on the main challenges to be met by these schemes, and we indicate possible good-practice lessons to be learned across instruments. The reader is referred to Chapter 5 for a detailed description of each instrument and to Chapter 6 for the results of each instrument's evaluation. The directions proposed below are seen as bases for practice-oriented benchmarking exercises involving policy-makers themselves, and, possibly, beneficiaries.

It should be noted here that such a view on innovation policy instruments is quite restrictive, since a wide range of other types of tools are acting on innovation behaviour: training support, investment support, fiscal and regulatory rules, environment regulations, competition policy etc. We focus here on the type of tools that have been subject to analyses in the SMEPOL study, without pretending that they cover the whole range of relevant policy instruments.

8.4.1 Direct-Support Schemes for R&D and Innovation Projects

Relevant instruments:
- FFF, ERP, ITF and RIP in Austria
- Development companies and Growth Fund in Denmark
- SMART in UK
- Recoverable advances in Wallonia
- Equipment loans in Italy
- National and regional grants or loans
- Tax deductions for private R&D investments.

Main challenges:
- Lower fragmentation of support according to various aspects of innovation process, take a longer-term view on support (ensure complementarity with awareness-raising, market-oriented, innovation management, commercialization, . . . support);
- Increase additionality (broaden and renew client base to pick up less obvious clients) without losing focus on innovation;
- Introduce more policy learning in these traditional policy tools;
- Work in complementarity with risk capital.

Good-practice lessons:
- NT has several good practice elements to offer to other schemes, combining high responsiveness and high co-ordination, and using an interactive mode of delivery of the support: high degree of policy learning, witnessed by incorporation of lessons from evaluations in subsequent programming periods, focus on learning on how to innovate in companies; all-round support covering financial, technical, commercial, managerial and organizational needs; long-term coaching of firms; policy implementers act as partners to the firms (presence on the board of companies); attention paid to foster linkages between firms and other agents etc.
- SMART could evolve towards incorporating more consideration on

marketing issues, notably by developing better co-ordination with Business Links;

- BIC takes equity in order to secure long-term support to companies (and payback);
- The Network of SMART winners introduce an inter-firm dimension in an otherwise very 'introverted' tool;
- Development Corporations in Denmark ensure a sparring partner function for supported firms, in addition to providing funds;
- More policy learning could be introduced in the recoverable advances scheme in Wallonia, though external evaluations, focusing on the analysis of reasons for success and failures in the supported projects.

8.4.2   Technical Personnel Introduction Schemes

Relevant instruments:
- KIM in Limburg
- RIT in Wallonia.

Main challenges:
- Increase penetration rate of the schemes;
- Increase additionality of the schemes, i.e. their role in changing behaviour in SMEs, rather than responding mainly to financial considerations;
- Upgrade flexibility of the schemes, to adapt them to firms' characteristics (nature of innovation process, level of formalization achieved etc.).

Good-practice lessons:
- The RIT scheme can usefully take lessons from the neighbouring KIM scheme and be transformed to: focus the selection criteria in order to ensure the introduction of a new function and new profile in the firm; extend the RIT support to marketing competences, and combine this scheme with the support of 'mentors' who 'sell' the scheme (helping firms expressing their needs) in order to ensure its success.

8.4.3   Technology Centres and Schemes Fostering Technological Diffusion to SMEs

Relevant instruments:
- GTS and TIC in Denmark
- Austrian Technology Centres

- Spanish Technology Centres
- RTC in UK
- Tecnopolis in Apulia and Service Centres in Lombardy
- TEFT and RUSH/REGINN in Norway.

Main challenges:
- Reconcile self-sufficiency on financial side with public service mission (awareness-raising in SMEs etc.);
- Combine role of developing supply side and respond to demands;
- Make the technology advisers evolve from a 'consultant' mode (transferring existing knowledge from their shelves to the firm) towards a 'process consultant' mode (working together with the firm on its transformation process, bringing in relevant knowledge – taken from anywhere – when necessary and adapting it to the particular situation). That is, developing a demand-led approach in these centres;
- Reconcile openness with context sensitivity: while proximity might help, it should not mean that support need necessarily be delivered from sources in proximity; a key role of a technology centre is to help firms find ways to relevant sources worldwide.

Good-practice lessons:
- TEFT's attachés perform a firm analyst function, which helps them evolve towards a more demand-oriented mission than if they were only in charge of transferring knowledge available in their centre to the firms;
- The Austrian centre Software Park, Hagenberg succeeds in performing a mix of the following functions: R&D, industrial development, teaching, incubation etc., which in the end favours interactive learning between researchers from the private and public sectors, enterprises managers, engineers and students;
- The Spanish Technology Centres are well embedded within the local industrial fabric, with entrepreneurs being present at conception and implementation of the services to be delivered by the centres (they have 'voice' in the policy). This could be an example to follow for the Tecnopolis centre in Apulia, for example, which suffers from a lack of linkages to business needs;
- The UK Lee Valley Centre works in co-operation with the Business Links, in order be able to offer a more complete support to firms involved in an innovation project, beyond their technological needs. This helps the former to evolve from a technology-led support towards more client-centred support;

- REGINN's approach is to support firm clusters rather than individual firms, which adds the 'interactive learning' dimension to the support.

8.4.4   Mobility Schemes for Researchers

Relevant instruments:
- FIRST-Enterprise in Wallonia
- FFF scheme for student stays in firms in Austria
- Mobility scheme in Denmark.

Main challenges:
- Focusing schemes on companies' needs rather than providing opportunities for financing research in the education sector;
- Ensuring a sufficient take-up of the scheme;
- Ensuring additionality of the scheme;
- Upgrading behavioural impact of the scheme in the longer-term, in terms of developing lasting collaborative patterns between research and industry.

Good-practice lessons:
- For the First-Enterprise scheme, extend the scheme to collaboration with other firms (rather than restricting it to public research laboratories) and to non-technological matters (managerial, strategic, marketing). Supplement this scheme with a new scheme for access to laboratories' equipment by SMEs.

8.4.5   Innovation Brokers and Innovation Advisers

Relevant instruments:
- Business Links and BICs in UK
- TIC in Denmark
- Syntens and KIC in Limburg
- TEFT in Norway.

Main challenges:
- Ensure a user-oriented mission while broker has a vested interest in the system;
- Professionalize such a job, often loosely defined; develop the skills of brokers in making tacit needs apparent;
- Improve the value-added to a 'pure' brokerage service;
- Maintain an innovation focus and avoid downgrading to 'basic' business development support;

- Reach micro-enterprises and less obvious clients;
- Achieve good linkages with other actors in the system;
- Demonstrate effectiveness of services with a long-term and rather uncertain impact.

Good-practice lessons:
- The 'fringe' tools in Wallonia (and other schemes) may learn from Business Links: they succeed in maintaining a holistic view on innovation and in carrying out a signposting function. The function of 'Personal Business Adviser' might be used as a basis to define more precisely the mentoring function developed on an ad hoc basis by several actors;
- Syntens evolved from pure brokers towards pro-active advisers, and added to their initial technological concern, competences in human resources development and strategic management. They also have been charged with a mission to reach new clients for their services;
- The quite innovative KIC scheme could provide lessons (on the positive and negative sides) to the numerous policy-makers interested in developing cluster policies in their region.

Although the above suggestions for schemes' improvement are presented per category of tools, one of the most promising approaches in the design of innovation policy lies in the development of tools that cross the boundaries of this typology: technology centres that also perform broker and innovation coach functions, direct financial support schemes that go along with support to human resources of innovation management, mobility schemes embedded in wider support to innovation projects etc. A move towards a more interactive, responsive and flexible innovation policy entails the development of multi-faceted instruments, or of strong linkages between traditionally distinct tools.

8.5. A SYNTHESIZED VIEW ON POLICIES DIRECTED AT INNOVATION IN SMEs IN A REGIONAL CONTEXT

The SMEPOL study has revealed a high degree of heterogeneity in policy instruments aiming at fostering innovation in SMEs. The instruments have various goals, such as linking SMEs with R&D-producing institutions or strengthening human resources within companies. The support also comes in various forms, like direct financial support, or services from technological centres or brokers, or under the name of cluster policy. Some policies clearly have a national origin while others may be classified as regional. Moreover, and perhaps of more fundamental importance, the policy instruments touch on different points of intervention in individual firms' innovation processes, or even different phases of the (collective) innovation path of the regional system these firms may belong to. For instance, the abilities and attitudes vis-à-vis innovation of firms targeted by awareness-raising instruments differ from the abilities and attitudes addressed by 'linear' instruments. Some tools might help to create the necessary awareness and capabilities in firms, so they can afterwards be supported with more standardized schemes when they have moved further up their learning curve. A proper sequence of instruments is then more appropriate than a search for universally and permanently adequate tools.

The evidenced diversity, of course, is not a surprise considering the variety among SMEs, their regional contexts and, most of all, the innovation processes. Notwithstanding this multilayered diversity, we can construct a simple two-dimensional classification of the policy instruments, which presents a synthesized view along two key issues concerning a shift or change in policy paradigm (see Table 8.1). We have classified the SMEPOL instruments according to the two following key fundamental characteristics:

1.   Target level of support: firm-oriented or (regional) system-oriented.

With the term 'system', we explicitly refer to regional systems. This does not imply that national or global systems or networks are irrelevant bases for economic co-ordination, but it expresses the claimed importance of the regional environment for innovation in SMEs. Some tools focus on innovation and learning within firms while others focus on crossing firm boundaries, aiming for externalities or synergies stemming from complementarity within the region as an innovation system. The logic behind (regional) system-oriented support is based on the idea that the innovation capacity and performance of a regional system may be larger than the 'sum' of the internal innovation capacity and performance of the individual 'members' of the system.

2.   Form and focus of support: focused on allocation of resources as inputs for innovation or focused on learning aiming for behavioural value-added.

At one extreme the policy approach is to raise the endowment, the stock of given resources (in firms and regions) as inputs for innovation. In a reactive mode of intervention, the policy instruments aim at increasing innovation capacity by making the necessary resource inputs available. The principal idea in the latter approach is that the window of opportunities and problems towards innovation and support are clear and that, given the lack (and need) of certain resource inputs, policy-makers increase the innovation output by allocating resources, that is, providing the innovation inputs or increasing their availability (again internally within the firm or externally, within the region). At the other extreme are the instruments which focus on learning, trying to change behavioural aspects like the organizational culture, innovation strategy, management, mentality or the level of awareness. They focus on creating or changing the windows of opportunities and problems concerning innovation and innovation policies. Accordingly, the mode of intervention is a proactive and interactive one. The principal idea is that the involved agents (private and public, individually or collectively) learn by doing, by using and by interacting. During innovation, using resources and interacting with others improves the awareness, the behavioural routines and the rationality towards innovation (and innovation policy).

Each of the four quadrants of Table 8.1 can be traced back to its own theoretical background or tradition ranging from atomistic to holistic approaches, and solutions from neo-classical and evolutionary traditions. The typology incorporates them all, and, in fact, it is suggested that in practice none of them is irrelevant in aiming for a change in innovative performance. 'Linear' tools directly aim for more innovation performance, while 'interactive' tools address innovation behaviour, but addressing behaviour is only meaningful if, in the end, it results in better performance.

Every policy in principle aims at changing behaviour. Policy-makers can affect the innovative behaviour of firms directly via subsidies and individual projects or indirectly via the provision of 'hard' or 'soft' public infrastructure and institutions like universities or a technology centre. In a neo-classical reasoning, providing subsidies as an input to the innovation process is an intervention method to affect the innovative behaviour of firms directly by influencing the choices based on the cost/benefit calculations of the agents. An input subsidy for R&D, or for hiring an expensive highly-educated employee for instance, affects the decisions regarding resource allocation immediately.

A more evolutionary approach to policy incorporates learning. In this respect subsidies for R&D can provide a learning experience. Within a 'learning-to-innovate' framework, policy support can get an innovation process started and support a change in the innovative behaviour in firms or regions. The support may also result in the static effect of more innovation output, but more importantly it aims for dynamic effects, effects which go on after the support stops. The argument for policy then becomes temporal.

Table 8.1 Classification of policy instruments studied in SMEPOL along mode and target of support

	Mode of innovation support	
Target level of support	Input resources (reactive tools allocating inputs for innovation)	Behavioural value-added (proactive tools focusing on learning to innovate)
Firm-oriented	**A**	**B**
	• *Traditional firms' R&D subsidies & loans*	• *Subsidy for hiring innovation managers in SMEs*
	• *Risk capital*	• *Loans for competence development*
	• *Training subsidies*	• *Incubators with 'soft' support*
	• *Incubators with 'hard' support*	• *Business Innovation Centres*
	• *Research centres*	• *'Proactive' technology centres*
	• *Traditional 'reactive' technology centres*	• *Audits, monitoring needs*
	• *Transfer units in universities*	• *Innovation Coach*
		• *Innovation management training & advice*
(regional) *System- oriented*	**C**	**D**
	• *Mobility schemes research industry*	• *Proactive Brokers, match-makers*
	• *Co-operative schemes HEI industry*	• *Cluster policies*
	• *Subsidy for co-operative R&D projects*	• *Support for firm networking*
	• *Subsidy to promote use of business services*	• *Umbrella schemes*
	• *Collective, User-oriented Technology or Innovation centres*	• *Local strategic plans*
		• *Schemes acting on the culture of innovation*
		• *RTP and RIS/RITTS kind of programmes*
		(fostering strategic capabilities of policy-makers)

Timing and the ex-ante conditions become important. The reasoning behind policy becomes proactive rather than re-active. The difficulty with reactive policy is to know exactly what the firm or the region needs. Some expressed needs may be 'over-supported' while others may be latent, neglected, tacit and not supported. The articulation of the need then has to become part of the policy process, albeit in an interactive way. Merely providing the resource inputs that the policy-makers think are relevant may not be enough to induce a real change in future behaviour.

An example of the difference between reactive and proactive tools can be provided by two instruments studied in SMEPOL, similar at first sight but serving different purposes, because one is more reactive and the other more proactive. Both the RIT scheme in Wallonia and the KIM scheme in Dutch Limburg, are subsidies to hire personnel in SMEs for the conduct of innovative projects, and have the objective of targeting firms that are not yet innovative. In the Walloon case, the firm itself has to write a formal technological development project and submit the proposal to the administration, who will decide on the subsidy according to the quality of the proposal and the results of an audit in the firm. The person employed needs to be a technician. In the Limburg case, an intermediary (Syntens, the innovation centre) helps the firm identify when such a scheme would be useful, helps find the candidate as needed, and does not require a formal project as a condition for the allocation of the subsidy. The type of personnel employed is not restricted to people possessing technological competences, but can also cover commercial or managerial weaknesses in the firm. It is clear that the RIT is mainly responding to the financial need of the company. Many SMEs do not use the RIT, because the formal requirement to codify a technological development project constitutes a barrier, and also because other financial sources are more easily accessible. In the case of KIM, the focus is more on the change of behaviour of the firm and there is an interaction between the firm and the support provider, within a more open view on the innovation project.

Table 8.1 can be used to examine under which paradigm issues (neoclassical or more evolutionary) the 40 SMEPOL policy instruments are developed and implemented. The A type of instruments may be classified as more 'traditional' while D type of instruments seem more 'innovative', but this does not mean that instruments in one of the four quadrants are intrinsically better than instruments in any of the other quadrants. There are still sound arguments for each and every type of tool. The question turns more into one of the choice of the appropriate policy portfolio, anticipating the needs of the region (see section 8.6 below). Concerning the resource-oriented tools A and C, the national policy level may in many cases be more relevant than the regional policy level, especially if the support is needed at firm level and the lack of (internal or external) resources for innovation is not region-specific.

The relevance of A, B, C as well as D types of tools not only relates to different regional conditions, like the intensity of existing co-operation practices, for instance, it also relates to the various identified distinctive characteristics of SMEs. The size-related characteristics affect the needs for support as well as the way the support can be delivered effectively. SMEs' limited resource base, for instance, finds a response in A or C types of tools. The A types focus on raising endowment within firms, and the C types of tools raise endowments of the innovation system the SME is part of, or is 'invited' to be part of. SMEs' distinctive organizational culture and management practices receive a better response in B types of tools, which try to influence certain attitudinal and behavioural aspects within the SMEs. The lesser ability of SMEs to shape their environment, compared to larger firms, might be addressed by D types of tools. These tools have to tackle the external uncertainties smaller firms are typically faced with, by enhancing the capacity of the firm to understand its environment and to become part of it (e.g. by forming clusters) – that is to become pro- and interactive members of a regional innovation system rather than staying passive, unaware and incapable of adapting or influencing others towards adaptation. This calls for external awareness, and tools which teach SMEs how to identify, value, use and develop regional resources and interact with resource owners.

The correspondence between the distinctive characteristics of SMEs regarding innovation and the various policy approaches shows, first of all, that each of the four types of policy instruments is relevant and, secondly, that the instruments have to be designed to address SMEs' needs, expressed as well as latent.

The main outcome of the SMEPOL study is that the policy tools are too concentrated in category A (e.g. in the form of firms' subsidies), and that there are few instruments in category D (e.g. support to cluster-forming). All types of instruments are relevant to different types of firms and different types of environments (and at different points in time), but the main gaps in the support systems in the SMEPOL cases are found in category D. In order to conduct a change in perspective, it seems necessary, however, in most cases to first develop instruments of the B and C types, before the system and agents are ready to implement and absorb D-type instruments.

As stated before, in any specific regional situation, there will probably be a need for a mix of A, B, C and D types of instruments. For example, A- and C-type instruments will be particularly relevant for New Technology-Based Firms and spin-offs. B-type instruments could be used for less innovation-aware firms. Building internal capabilities is a necessary step in most cases before being able to participate in a D-type instrument, promoting interaction with other innovating agents.

If a region does not have a lot of innovative SMEs, providing ever more resources to the same group of firms seems less appropriate than extending the number of innovators by approaching non-innovating SMEs with B-type of tools. Enhancing their learning process and preparing them for more interactive behaviour can subsequently be supported by C or D types of tools. If there is no lack of innovators but they seem to innovate in relative isolation, C types of tools might create more openness and stimulate the use of external resources in the region.

There are also arguments to develop linkages between tools of the various categories. In general, however, the proactive provision of internal and external learning experiences with B and D types of instruments respectively will create new clients, and new resource needs which may subsequently be effectively addressed with the reactive provision of internal or external resources. The other logical sequential link refers to the fact that a certain level of internal resources and learning experiences are needed before system-oriented tools can be effective. This calls for excellent co-ordination and the development of synergies between all tools at work in the environment.

This discussion also links with the question of the relevance of different levels of authorities for the various policy activities like design, adaptation, learning, implementation and evaluation. Proactive tools imply more freedom of action and closeness to beneficiaries more likely to be found at regional level, while reactive, standard tools are more adapted to higher levels of authorities.

8.6. CUSTOMIZING A POLICY PORTFOLIO TO REGIONAL SPECIFICITIES

An argued in several chapters of this book, regional differences in innovation capabilities call for a tailored mix of policy instruments. This section, therefore, reflects on the question of the appropriate policy portfolio to be developed in a regional context. For targeting of policy, characteristics of the region as a whole and SME-specific challenges for innovation are taken as the point of departure. To achieve the objective of identification of policy portfolios, Table 8.2 proposes a picture of the situation which combines the results of the analyses gathered in the SMEPOL study, i.e. the analysis of the main potential deficits of regional innovation systems (see Chapter 3), and the analysis of the main firms' barriers with regard to innovation (see Chapters 1 and 2).

Table 8.2 *Policy responses to regions' and SMEs' problems with innovation: an illustration of possible policy objectives and tools*

| SMEs' needs | Regional innovation system problems | | |
	Organizational thinness	Fragmentation	Lock-in
Financial	*Attract and retain innovating firms* *Foster firms to go global, link them to international partners and external financial resources*	*Coach firms in linking to finance sources* *Foster specialization by combining technological support and finance* *Support the formation of sector-specific venture-capital funds*	*Ensure long-term finance for 'overall' innovation project* *Support new firm creation* *Address succession problems* *Restructure mature industries by attracting FDI and promoting international partnerships*
Technological	*Link firms with technological resources outside the region* *Promote collective sourcing and investments in collective resources* Technology centres *Use private R&D centres as technology resources for other firms*	*Provide bridge between firms and technological resources, brokers* *Provide 'accessible' technology: co-operative schemes HEI-SMEs* *Finance firms to access technology centres*	*Push firms to seek new technology options* *Foster access to resources outside the region, international partnerships* *Restructure the technology support infrastructure towards new technologies and sectors*
Human resources	*Attract/retain highly-skilled workers* *Support collective training programmes*	*Foster exchange of codified and tacit knowledge* *Foster intra-firm nodes for co-operation: hiring of researchers in SMEs*	*Develop creative capacities of workers* *Subsidy to hire personnel for innovation*

Openness and learning attitude	*Promotion of networking between firms, and clusters at every geographical scale*	*Foster a more collaborative spirit and more strategic orientation in the regions*, Regional Development Agencies *Combine funding with interaction stimulation*, Umbrella schemes	*Help SMEs evolve towards more creativity and autonomy in production* Supply-chain learning, Demonstration projects
Strategy & Organization	*Support firms in linking to international input and output markets* *Develop systemic character of innovation support*: network brokers	*Help firms identify, articulate and 'de-bundle' their needs*: innovation coach *Invite firms collectively to help in formulating a regional innovation strategy*: RIS/RITTS	*Open windows of opportunities for SMEs: techno-economic intelligence schemes* Innovation management training

The combination of regional and firms' deficits should form the basis for the design of policy intervention. The aim of Table 8.2 is to enlighten possible policy responses to certain innovation barriers, deficits or challenges at regional and firm level. Such a table shows clearly that there is no 'one-size-fits-all' policy system: it depends on the problems and opportunities to be addressed in the existing context. It also shows the need for policy to provide longer term and holistic support to innovation in all its aspects. But it is nevertheless possible to develop recommendations per category of firms' problems (horizontal lines in the table), per regional context (vertical lines in the table), and per type of policy tool (within each cell, every tool can be benchmarked against the broad guidelines for policy and against similar tools of the same type). In light of such a table of possible instruments, it becomes clear that the main role of the policy-maker becomes setting priorities according to the perceived problems in the region, to align with the main orientation of the economic development policy of the region.

The main condition however for the usefulness of the proposed approach, as a guide to re-designing innovation policies targeting SMEs, is that on the policy side too, learning processes are at stake, and adequate strategic capabilities are present. In trying to create learning firms within learning regions there is a need for learning governments. This implies four things in particular:

1. That the regional situation, that is the particular needs and opportunities (for innovation support) of SMEs and the regional innovation system as a whole, are well mapped, communicated and understood by policy designers;
2. That the objectives set for policy instruments are clearly expressed ex-ante in a global coherent framework and that the expected results are measurable;
3. That the results and impacts of the instruments are monitored properly and then evaluated by an independent third party in conjunction with peers' and clients' views;
4. That lessons from the evaluation are acknowledged and diffused and that they are compared with the policy aims, in order to adjust the policy approach in a continuous manner, and its instruments accordingly.

In most of the regions studied in SMEPOL, deficiencies have been noted in all the aspects listed above: a detailed knowledge of the specificities of the regional innovation system is often absent, not properly diffused in a language understandable by policy-makers, or not updated with sufficient regularity; effects of policies are often measured in a 'funds-consumption' approach only; evaluation is rarely built into the design of the programmes;

no real independent evaluation of results and impacts are undertaken in most cases; pilot, bottom-up experiments are not really assessed, and there is thus a failure to capture lessons from these experiments; and policy learning is, in general, underdeveloped.

8.7 CONCLUSION

Drawing on the SMEPOL findings we have questioned in this chapter how policy directed at innovation in SMEs can be improved. After discussing several rationales for policy we came up with a main question, i.e. did we witness an actual shift in the policy paradigm or did we find arguments which call for a new policy paradigm? Our conclusion is that we have witnessed some shifts in practice, and that we have found sound arguments which support our claim that a shift in rationale is needed to improve the policy directed at innovation in SMEs.

More precisely, we used Table 8.1 to discuss the way to combine tools that are reactive or proactive and target internal processes in SMEs, and tools that are proactive and target the externalization of SMEs or the functioning of the regional innovation system.

We also proposed a reflection on policy mixes, using Table 8.2, showing that similar SME problems need to be tackled differently according to the regional context, but also that there is scope for importing elements of good practices from one context into another. A key challenge for innovation policy lies in the development of appropriate complementarities between instruments, in order to address the diversity of modes of innovation among SMEs and of innovation barriers in regions. This is an area where much progress is still to be achieved.

With this discussion, we can conclude that such an analysis, with the three key features of:

- Matching the context and SME needs' pictures with the policy tools in each region
- Confronting the policy tools with the lessons of theory
- Comparing results achieved with a range of policy instruments in different environments

is suitable for improving SME innovation-support policies in the EU regions.

The best way to evolve in such a direction, however, would be to undertake such a strategic benchmarking exercise with the active participation of policy-makers and policy implementers themselves. Theoretically sensible ideas could then be confronted with reality.

Notes

1. Camagni (1991) offers a related perspective to explain how firms translate external information into a language that the firm can understand via a 'transcoding function'. According to Camagni: 'These functions are perhaps the most critical, though widely overlooked by economic theory, in that they control the process of interfirm know-how transfer and information appropriation. Utilizing codified information, both freely available or costly, and merging it with chaotic and unordered "information" results in a firm-specific "knowledge" and possibly into potential business ideas at the disposal of the managerial decision-making' (Camagni, 1991, p.127).
2. A third type of coordinated economies (not further discussed in the chapter) is the keiretsu- or chaebol-type coordination among groups of companies in industries in Japan and Korea.
3. Part of this chapter was first published in Musyck, B. and G. Garofoli (2001), 'Innovation policies for SMEs in Europe: towards an interactive model?', *Regional Studies*, 35(9), pp. 869–72. Reprinted with kind permission of the Taylor and Francis Group (www.tandf.co.uk).
4. The additionality of the policy tools is related to the changes that would not have occurred without its implementation.
5. LVBIC from the United Kingdom could also be included in this section.
6. BI, LVCs and LVBIC from the United Kingdom could also be included in this section.
7. As an example the UK team found that some technological centres are catering for the needs of large firms as well as SMEs in their study of Regional Technology Initiatives (RTIs) in the English regions (Smallbone et al. 1999).
8. The LVCs from the United Kingdom could also be included in this section.
9. A previous version of this chapter has been published in the *Technology Analysis and Strategic Management Journal* (Nauwelaers and Wintjes 2002). The authors wish to thank their colleague Viki Sonntag for her valuable comments on an earlier draft of this paper.
10. Even there, in most cases, the problem is not so much the existence of risk capital funds but the accessibility of these for SMEs engaged in risky, and sometimes weakly formalized projects.

Bibliography

Acs, Z.J. and D.B. Audretsch (1990), *Innovation and Small Firms*, Cambridge, MA: MIT Press.

Acs, Z.J. and A. Varga (2002): 'Geography, endogenous growth, and innovation', *International Regional Science Review*, **25** (1), 132–48.

Amin, A. and P. Cohendet (1999), 'Learning and adaptation in decentralised business networks', *Environment and Planning D: Society and Space*, 17, 87–104.

Amin, A. (1999), 'An institutionalist perspective on regional development', *International Journal of Urban and Regional Research*, **2** (2), 365–78.

Amin, A. and K. Robins (1991), 'These are not Marshallian times', in R. Camagni (ed.), *Innovation Networks: Spatial Perspectives*, London/New York: Belhaven Press, pp. 105–18.

Amin, A. and N. Thrift (1994), 'Living in the global', in A. Amin and N. Thrift (eds), *Globalization, Institutions and Regional Development in Europe*, Oxford: Oxford University Press, pp. 1–22.

Amin, A. and N. Thrift (1995), 'Territoriality in the global political economy', *Nordisk Samhällsgeografisk Tidskrift*, **20**, 3–16.

Andersen, P.H. and P. R. Christensen (1998), *Den globale udfordring – Danske underleverandørers internationalisering*, Copenhagen: Erhvervsfremme Styrelsen.

Archibugi, D. and J. Michie (1995), 'Technology and innovation. An introduction', *Cambridge Journal of Economics*, 19, 1–4.

Argyris, C. and D. Schon (1978), *Organisational Learning*, London: Addison-Wesley.

Asheim, B.T. (1994), 'Industrial districts, inter-firm co-operation and endogenous technological development: the experience of developed countries', in *Technological Dynamism in Industrial Districts: An Alternative Approach to Industrialization in Developing Countries?*, New York and Geneva: UNCTAD, pp. 91–142.

Asheim, B.T. (1996), 'Industrial districts as "learning regions": A condition for prosperity?', *European Planning Studies*, 4 (4), 379–400.

Asheim, B.T. (1999), 'Innovation, social capital and regional clusters: On the importance of co-operation, interactive learning and localised knowledge in learning economies', paper presented at the European Regional Science Association 39th European Congress, Dublin, Ireland, 23–27 August.

Asheim B.T. (1999b), 'The territorial challenge to innovation policy: agglomeration effects and regional innovation systems', in B. Asheim and K. Smith (eds), *Regional Innovation Systems, Regional Networks and Regional Policy*, Cheltenham, UK and Lyme, US: Edward Elgar.

Asheim, B.T. (2001), 'Learning regions as development coalitions: partnership as governance in European welfare states? Concepts and transformation', *International Journal of Action Research and Organizational Renewal*, **6**(1), 73–101.

Asheim, B.T. and P. Cooke (1998), 'Localised innovation networks in a global economy: A comparative analysis of endogenous and exogenous regional development approaches', *Comparative Social Research*, **17**, 199–240.

Asheim, B.T. and P. Cooke (1999), 'Local learning and interactive innovation networks in a global economy', in E. Malecki and P. Oinas (eds), *Making Connections. Technological Learning and Regional Economic Change*, Aldershot: Ashgate, 145–78.

Asheim, B.T. and A. Isaksen (1997), 'Location, agglomeration and innovation: Towards regional innovation systems in Norway', *European Planning Studies*, **5** (3), 299–330.

Asheim, B.T and A. Isaksen (2000), 'Localised knowledge, interactive learning and innovation: between regional networks and global corporations', in E. Vatne and M. Taylor (eds), *The Networked Firm in a Global World. Small Firms in New Environments*, Aldershot: Ashgate, pp. 163–98.

Autio, E. (1998), 'Evaluation of RTD in regional systems of innovation', *European Planning Studies*, **6** (2), 131–40.

Bank of England (1996), *The Financing of Technology-Based Small Firms*, London: Bank of England.

Becattini G. (ed.) (1987), *Mercato e forze locali: il distretto industriale*, Bologna: Il Mulino.

Becattini G. (ed.) (1989), *Modelli locali di sviluppo*, Il Mulino, Bologna.

Becattini G. (1998), *Distretti industriali e made in Italy*, Turin: Bollati Boringhieri.

Bellandi, M. (1994), 'Decentralised industrial creativity in dynamic industrial districts', in *Technological Dynamism in Industrial Districts: An Alternative Approach to Industrialization in Developing Countries?*, New York and Geneva: UNCTAD, pp. 91–142.

Bessant J., S. Caffyn, J. Gilbert, R. Harding, and S. Webb (1994), 'Rediscovering continuous improvement', *Technovation*, **14** (1), 17–29.

Bessant, J., S. Caffyn, and S. Austin (1995) 'Continuous improvement and organisational learning', paper presented at the R&D Management Conference, Pisa, Italy, 20–22 September.

Birch, D. (1979), *The Job Generation Process*, Cambridge, MA: MIT Press.

Bonaccorsi, A. and A. Lipparini (1994), 'Strategic partnerships in new product development: An Italian case study', *Journal of Product Innovation Management*, **11** (2), 135–46.

Bower, G. H. and E. R. Hildegard (1981), *Theories of Learning*, Engelwood Cliffs, NJ: Prentice-Hall.

Braczyk, H.-J., P. Cooke and M. Heidenreich (eds) (1998), *Regional Innovation Systems*, London: UCL Press.

Brusco, S. (1982), 'The Emilian model: productive decentralisation and social integration', *Cambridge Journal of Economics*, **6** (2) 47–62.

Brusco S. (1989), *Piccole imprese e distretti industriali*, Turin: Rosenberg & Sellier.

Camagni, R. (1991), *Innovation Networks: Spatial Perspectives*, London: Belhaven-Pinter.

Cappellin, R. and M. Steiner (2002), 'Enlarging the scale of knowledge in innovation networks: theoretical perspectives and policy issues', paper presented to the 42nd Congress of the European Regional Science Association, Dortmund, August 27–31.

Capron, H. and W. Meeusen (1999), 'National innovation systems. Pilot study of the Belgian innovation system', study carried out for the Belgian Federal Office for Scientific, Technical and Cultural Affairs (OSTC) in the context of the OECD Working Group on Innovation and Technology policy.

Carson, D. (1991), 'Research into small business marketing', *European Journal of Marketing*, **9**, 75–91.

Castro, E. de and C. Jensen-Butler (1993), 'Flexibility, routine behaviour and the neo-classical model in the analysis of regional growth', working paper of the Department of Political Science, University of Aarhus, Aarhus, Denmark.

CEC (1987), *Job Creation in Small and Medium Enterprises, Programme of Research and Actions on the Development of the Labour Market: Summary Report*, Brussels: Commission of the European Communities.

Christensen, J.F. (1996), *Teknologisk Service: Tendenser og udfordringer. En diskussion af GTS-Institutternes værdi for Danmark*, Copenhagen: Institutrådet.

Christensen, P. R., Cornett, A., and K. Philipsen (1999), *Innovations and Innovation Support for SMEs. The Triangle Region of Denmark*, SMEPOL report no. 2, Kolding: Centre for Small Business Research. Southern Denmark University.

Cobbenhagen, J. (1999), *Managing Innovation at Company Level*, Maastricht: Universitaire Pers Maastricht.

Cobbenhagen, J., F. den Hertog, and H. Pennings (1995), *Succesvol veranderen*: *kerncompetencies and bedrijfsvernieuwing*, The Hague: Kluwer Bedrijfswetenschappen.

Cohen, W.M. and D.A. Levinthal (1990), 'Absorptive capacity: a new perspective on learning and innovation', *Administrative Science Quarterly*, **35**, 128–52.

Cooke, P. (1994), 'The co-operative advantage of regions', paper presented for the conference on 'Regions, institutions, and technology: Reorganizing economic geography in Canada and the Anglo-American World', University of Toronto, September.

Cooke, P. (1995), 'Planet Europe: network approaches to regional innovation and technology management', *Technology Management*, 2, 18–30.

Cooke, P. (1998), 'Introduction. Origins of the concept', in H.-J. Braczyk et al. (eds), *Regional Innovation Systems*, London: UCL Press, pp. 2–25.

Cooke, P. and K. Morgan (1993), 'The network paradigm – new departures in corporate and regional development', *Environment & Planning D: Society and Space*, **11**.

Cooke, P. and K. Morgan (1998), *The Associational Economy. Firms, Regions, and Innovations*, Oxford: Oxford University Press.

Cooke, P., P. Boekholt and F. Tödtling (eds) (2000), *The Governance of Innovation in Europe – Regional Perspectives and Global Competitiveness*, London and New York: Pinter.

Cosh, A. and A. Hughes (1994), 'Size, financial structure and profitability: UK Companies in the 1980s' in A. Hughes and D. Storey (eds), *Finance and the Small Firm*, London: Routledge.

Cosh, A., J. Duncan and E. Wood (1996), 'Innovation: scale objectives, and constraints', in A. Cosh and A. Hughes (eds), *The Changing State of British Enterprise; Growth Innovation and Competitive Advantage in SMEs 1986–95*, Cambridge: ESRC Centre for Business Research, University of Cambridge.

Cowan, R. and D. Foray (1997), 'The economics of codification and the diffusion of knowledge', *Industrial and Corporate Change*, **6** (3), 595–622.

Cowan, R. and G. van de Paal (2000), *Innovation Policy in a Knowledge-based Economy*, Luxembourg: Publication N° EUR 17023 of the European Commission.

Craggs, A. and P. Jones (1998), 'UK results from the Community Innovation Survey', *Economic Trends*, **539** (October), 51–7.

Crevoisier, O. (1994), 'Book review of Benko, G. and A. Lipietz (eds), Les régions qui gagnent, Paris 1992', *European Planning Studies*, 2, 258–60.

Curran, J. (1993) 'TECs and small firms: can TECs reach the small firms other strategies have failed to reach?', paper presented to the All Party Social Science Group, London: House of Commons.

Curran J. and R. Blackburn (1994), *Small Firms and Local Economic Networks: the Death of the Local Economy?*, London: Paul Chapman Publishing.

Cyert, R. M. and J. G. March (1963), *A Behavioural Theory of the Firm*, Engelwood Cliffs, NJ: Prentice-Hall.

Dankbaar, B. (1998), 'Technology management in technology-contingent SMEs', *International Journal of Technology Management*, 15 (1/2), 70–81.

Davis, A. (1995), 'Local economies and globalisation', internal paper, Paris: OECD.

Davis S.J. and J. Haltiwanger (1992), 'Gross job generation, gross job destruction and employment re-allocation', *Quarterly Journal of Economics*, **107**, 819–63.

DeBresson, C. and R. Walker (eds) (1991), 'Network of innovators', *Research Policy* (special issue), **20**, 5.

DiMaggio, P.J. and W.W. Powell (1983), 'The iron cage revisited: institutional isomorphism and collective rationality in organisational fields', *American Sociological Review*, **48**, 147–60.

Dosi, G. (1988), 'The nature of the innovative process', in: G. Dosi, Ch. Freeman, R. Nelson, G. Silverberg and L. Soete (eds), *Technical Change and Economic Theory*, London: Pinter Publishers, pp. 221–38.

EC (1995), *Green Paper on Innovation*, Luxembourg: Bulletin of the European Union, Supplement 5/95.

EC (1999), *Second European Report on S&T Indicators 1997*, Luxembourg: European Commission, EUR 17639.

Edquist, C. (ed.) (1997), *Systems of Innovation – Technologies, Institutions and Organizations*, London/Washington: Pinter.

Edquist, C. and B.-Å. Lundvall (1993), 'Comparing the Danish and Swedish systems of innovation', in R. R. Nelson (1993) (ed), *National Innovation Systems. A Comparative Analysis*, New York, Oxford: Oxford University Press, pp. 265–98.

EIM (1994), *The European Observatory for SMEs: Second Annual Report*, Zoetermeer, NL: EIM Small Business Research and Consultancy.

EIM (1997), *The European Observatory for SMEs: Fifth Annual Report*, Zoetermeer, NL: EIM Small Business Research and Consultancy.

Ellis, H. C. (1965), *The Transfer of Learning*, New York: Macmillan.

Encyclopædia Britannica (1999), International version (CD).

Ennals, R. and B. Gustavsen (eds) (1999), *Work organisation and Europe as development coalition*, Amsterdam/Philadelphia: John Benjamin's Publishing Company.

Ernst, D, P. Guerrieri, S. Iammarino and C. Pietrobelli (2001), 'New challenges for industrial clusters and districts: global production networks and knowledge diffusion', in P. Guerrieri, S. Iammarino and C. Pietrobelli (eds), *The Global Challenge to Industrial Districts*, Cheltenham, UK and Northampton, USA: Edward Elgar, pp. 131–44.

Estes, W. K. (1970), *Learning Theory and Mental Development*, New York: Academic Press.

Eurostat (1998), *Community Innovation Survey 1997/1998,* Luxembourg.

Eurostat (2001), *Statistics on Innovation in Europe. Data 1996–1997*, Luxembourg: European Communities.

Feldman, M. (1994), *The Geography of Innovation*, Dordrecht: Kluwer Academic Publishers.

Fendel, R. and M. Frenkel (1998), 'Do small and medium-sized enterprises stabilise employment?' in *Zeitschrift für Wirtschafts- und Sozialwissenschaften (ZWS)*, **118** (2), 163–84.

Freeman, C. (1993), *The political economy of the long wave*, Paper presented at EAPE 1993 conference on 'The economy of the future: ecology, technology, institutions', Barcelona, October.

Freeman, C. (1995), 'The "national system of innovation" in historical perspective', *Cambridge Journal of Economics*, **19**, 5–24.

Gaffard, J. C. (1992), *Territory as a Specific Resource: the Process of Construction of Local Systems of Innovation*, Nice: Latapses.

Garnsey, E. and I. Moore (1993), 'Pre-competitive and near market research and development: problems for innovation policy', in M. Dodgson and R. Rothwell (eds), *Small Firms and Innovation: The External Influences,* a special publication of the *Journal of Technology Management*, 69–83.

Garofoli, G. (1983), *Industrializzazione diffusa in Lombardia*, Milan: Franco Angeli Editore, (2nd edn Pavia: Iuculano, 1995).

Garofoli, G. (1989), 'Industrial districts: structure and transformation', *Economic Notes*, **1**, 37–54.

Garofoli G. (1991), *Modelli locali di sviluppo*, Milan: Franco Angeli Editore.

Garofoli, G. (1992) (ed.), *Endogenous Development and Southern Europe*, Aldershot: Avebury.

Garofoli, G. (ed.) (1999), *SMEs, Innovation Trajectories and Policies: the Case of Lombardy and Apulia. SMEPOL Report no. 3*, Pavia: Dipartemento di Economia Politica e Metodi Quantitative, Università degli Studi di Pavia.

Garofoli, G. (1999b), 'Sistemi locali di impresa e performance dell'impresa minore in Italia', in F. Traù (ed.), *La questione dimensionale nell'industria italiana*, Bologna: Il Mulino.

Glasmeier, A. (1994), 'Flexible districts, flexible regions? The institutional and cultural limits to districts in an era of globalisation and technological paradigm shifts', in A. Amin and N. Thrift (eds), *Globalization,*

Institutions and Regional Development in Europe, Oxford: Oxford University Press, pp. 118–46.

Gottardi, G. (1996), 'Technology strategies, innovation without R&D and the creation of knowledge within industrial districts', *Journal of Industry Studies*, 3, 119–34.

Grabher, G. (1993a), *The Embedded Firm – on the Socio Economics of Industrial networks*, London: Routledge.

Grabher, G. (1993b), 'Rediscovering the social in the economics of inter-firm relations', in G. Grabher, *The Embedded Firm – on the Socio Economics of Industrial Networks*, London: Routledge.

Granovetter, M (1973), 'The strength of weak ties', *American Journal of Sociology*, **78** (6), 60–80.

Granovetter, M. (1985), 'Economic action and social structure: the problem of embeddedness', *American Journal of Sociology*, **91** (3) 481–510.

Gregersen, B. and B. Johnson (1997), 'Learning economies, innovation systems and European integration', *Regional Studies*, **31** (5) 479–90.

Håkansson, H. (1992), *Corporate Technological Behaviour. Co-operation and Networks*, London: Routledge.

Haraldsen, T. (1995), 'Spatial conquest – the territorial extension of production systems,' paper presented at the Regional Studies Association conference on 'Regional Futures: Past and Present, East and West', Gothenburg, May.

Hart, D. and J. Simmie (1997) 'Innovation, competition and the structure of local production networks: initial findings from the Hertfordshire Project', *Local Economy*, November.

Hassink, R. (1996), 'Technology transfer agencies and regional economic development', *European Planning Studies*, **4**(2), 167–84.

Hawkins, P. (1994), 'The changing view of learning', in J. Burgoyne, M. Pedlar and T. Boydell (eds), *Towards the Learning Company: Concepts and Practices*, New York: McGraw-Hill, pp. 9–27.

Hodgson, G.M. (2002), 'Introduction', in G.M. Hodgson, (ed.), *A Modern Reader in Institutional and Evolutionary Economics*, Cheltenham, UK and Northampton, USA: Edward Elgar, pp. xiii–xxix.

Hoffman, K., M. Parejo, J. Bessant and L. Perren (1998) 'Small firms, R&D, technology and innovation in the UK: a literature review', *Technovation*, **18**(1), 39–55.

Holmes, J. (1986), 'The organization and locational structure of production subcontracting', in A. Scott and M. Storper (eds), *Production, Work, Territory*, Boston, MA: Allen & Unwin, pp. 80–106.

Howells, J. (1996), 'Regional systems of innovation?', paper presented at HCM conference on 'National systems of innovation or the globalisation of technology? Lessons for the public and business sector', ISRDS-CNR, Rome, April.

Isaksen, A. and S.O. Remøe (2001), 'New approaches to innovation policy: some Norwegian examples', *European Planning Studies*, **9** (3), 285–302.

Isaksen, A., B. Asheim and S.O. Remøe (1999) (eds), *SME policy and the regional dimension of innovation. The Norwegian report. SMEPOL report no. 5*, Oslo: STEP Group.

Kaufmann, A. and F. Tödtling (1999), *Innovation Support for SMEs in Upper Austria. SMEPOL report no. 1*, Vienna: Institute for Urban and Regional Studies. Vienna University of Economics and Business Administration.

Keeble D., C. Lawson, H. Lawton-Smith, B. Moore, and F. Wilkinson (1997), 'Internationalisation processes, networking and local embeddedness in technology-intensive small firms', in M. Ram, D. Deakins and D. Smallbone (eds), *Small Firms: Enterprising Futures*, London: Paul Chapman Publishing.

Kelly, G.A. (1955), *The Psychology of Personal Constructs*, 1 & 2, New York: Norton.

Kitschelt, H., P. Lange, G. Marks and J.D. Stephens (1999), 'Convergence and divergence in advanced capitalist democracies', in H. Kitschelt, P. Lange, G. Marks and J.D. Stephens (eds), *Continuity and Change in Compemporary Capitalism*, Cambridge: Cambridge University Press, pp. 427–60.

Kline, S. J. and N. Rosenberg (1986), 'An Overview of Innovation', in R. Landau and N. Rosenberg (eds), *The Positive Sum Strategy*, Washington: National Academy Press.

Koschatzky, K. (1999), 'National versus regional systems of innovations – Crossborder linkages and learning between Baden-Württemberg and Alsace', paper presented at the NECSTS-99 Conference on 'Regional Innovation Systems in Europe', San Sebastian, Spain, 30 September–2 October.

Krugman, P. (1991), *Geography and Trade*, Leuven: Leuven University Press.

Lageman, B. and K. Lobbe (1999), *Kleine und Mittlere Unternehmen im Sektoralen Strukturwandel*, Essen: Untersuchungen des RWI, No. 28.

Landabaso, M. (1997), 'The promotion of innovation in regional policy: proposals for a regional innovation strategy', *Entrepreneurship and Regional Development*, **9**, 47–65.

Lane, P. J. and M. Lubatkin (1998), 'Relative absorptive capacity and interorganisational learning', *Strategic Management Journal*, **19**, 461–77.

Lazonick, W. (1993), 'Industry cluster versus global webs: organizational capabilities in the American economy', *Industrial and Corporate Change*, **2**, 1–24.

Lazonick, W. and M. O'Sullivan (1996), 'Sustained economic development', *STEP Report*, 14.

Leborgne, D. and A. Lipietz (1988), 'New technologies, new modes of regulation: some spatial implications', *Environment and Planning D: Society and Space*, **6**, 263–80.

Leborgne, D. and A. Lipietz (1992), 'Conceptual fallacies and open questions on post-Fordism', in M. Storper and A.J. Scott (eds), *Pathways to Industrialization and Regional Development*, London: Routledge, pp. 332–48.

Leborgne D. and A. Lipietz (1992b), 'Flexibilité offensive, flexibilité défensive. Deux stratégies sociales dans la production des nouveaux espaces économiques', in G. Benko and A. Lipietz (eds), *Les Régions qui gagnent. Districts et réseaux: les nouveaux paradigmes de la géographie économique*, Paris: Presses Universitaires de France.

Lundquist, L. (1974), *Förvaltningen i det politiska systemet*, Lund: Studenterlitteratur.

Lundvall, B.-Å. (1983), 'User–producer relationships, national systems of innovation and internationalisation', in D. Foray and Freeman, C. (eds), *Technology and the Wealth of Nations*, London: Pinter Publishers.

Lundvall, B.-Å. (1992), 'Introduction', in B.-Å. Lundvall (ed.), *National Systems of Innovation*, London: Pinter Publishers, pp. 1–19.

Lundvall, B.-Å. (1996), 'The social dimension of the learning economy', *DRUID Working Papers*, **96** (1) 1–45.

Lundvall, B.-Å. and S. Borrás (1997), *The Globalising Learning Economy: Implications for Innovation Policy*, Luxembourg: Office for Official Publications of the European Communities.

Lundvall, B.-Å. and B. Johnson (1994), 'The learning economy', *Journal of Industry Studies*, **1** (2), 23–42.

Malecki, E. (1997), *Technology and Economic Development: The Dynamics of Local, Regional and National Competitiveness*, 2nd edn, London: Longman, Addison Wesley.

Malecki, E. and P. Oinas (1999), 'Spatial innovation systems', in E. Malecki and P. Oinas (eds), *Making Connections – Technological Learning and Regional Economic Change*, Aldershot: Ashgate.

Malerba, F. (1993), 'The national systems of innovation: Italy', in R. Nelson, (1993) (ed.), *National Innovation Systems. A Comparative Analysis*, New York, Oxford: Oxford University Press, pp. 230–59.

Malmberg, A. (1997), 'Industrial geography: location and learning', *Progress in Human Geography*, **21** (4), 573–82.

Malmberg, A. and P. Maskell (1999), 'Guest editorial: localized learning and regional economic development', *European Urban and Regional Studies*, **6** (1), 5–8.

March, J.G. and H.A. Simon (1958), *Organizations*, New York: Wiley.

Mariussen, Å. (1997), *Immobil kunnskap og globale innovasjonssystem*, Bodø: Nordland Research.

Markusen, A. (1996), 'Sticky places in slippery space: a typology of industrial districts', *Economic Geography*, **72** (3), 293–313.

Maskell, P. and A. Malmberg (1998), *Competitiveness, Localised Learning and Regional Development. Specialisation and Prosperity in Small Open Economies*, London and New York: Routledge.

McKinsey and Company (1987), *Genereren, overdragen en toepassen van technologishe kennis*, Amsterdam: McKinsey and Company – cited in Hassink (1996).

Mintzberg, H. and J. A. Waters (1985), 'Of strategies, deliberate and emergent', *Strategic Management Journal*, 257–72.

Mintzberg, H., J. B. Quinn and S. Goshal (1995), *The Strategy Process. European Edition*, London: Prentice-Hall.

Morgan, K. (1997), 'The learning region: institutions, innovation and regional renewal', *Regional Studies*, **31** (5), 491–503.

Morgan, K. and C. Nauwelaers (1999), *Regional Innovation Strategies: the Challenge for Less-Favoured Regions*, London: The Stationery Office and The Regional Studies Association, Taylor and Francis Group, Routledge.

Nauwelaers, C. (1997), 'Evaluating regional innovation potential: assessment of trends and implications for policy', in J. Mitra and P. Formica (eds), *Innovation and Economic Development*, Dublin: Oak Tree Press, pp. 217–33.

Nauwelaers, C. (2000), *Policy learning for innovation in European regions*, paper presented at the RESTPOR conference, Kashikojima, Japan, September 5–7.

Nauwelaers, C. (2001), 'Path dependency and the role of institutions in cluster policy generation', in A. Mariussen (ed.), *Cluster Policies–Cluster Development?*, Stockholm: Nordregio Series, 2, 93–107.

Nauwelaers, C. and K. Morgan (1999), 'The new wave of innovation-oriented regional policies: retrospects and prospects', in K. Morgan and C. Nauwelaers (1999), *Regional Innovation Strategies: the Challenge for Less-Favoured Regions*, London: The Stationery Office and The Regional Studies Association, Taylor and Francis Group, Routledge, pp. 224–38.

Nauwelaers, C. and R. Wintjes (2002), 'Innovating SMEs and regions: the need for policy intelligence and interactive policies', *Technology Analysis and Strategic Management*, **14** (2), 201–15.

Nauwelaers, C., R. Wintjes and N. Schall (1999), *SME Policy and the*

Regional Dimension of Innovation: The cases of Wallonia and Limburg.
SMEPOL report no. 4, Maastricht: MERIT, Maastricht University.

Nelson, R. (1993), 'A retrospective', in R. Nelson (ed.), *National Innovation Systems. A Comparative Analysis*, New York, Oxford: Oxford University Press, pp. 505–23.

Nelson, R. and N. Rosenberg (1993), 'Technical innovation and national systems', in R. Nelson (ed.), *National Innovation Systems. A Comparative Analysis*, New York, Oxford: Oxford University Press, pp. 3–21.

Nelson, R. and S. Winter (1982), *An Evolutionary Theory of Economic Change*, Cambridge, MA: Harvard University Press.

Nonaka, I. and P. Reinmöller (1998), 'The legacy of learning. Toward endogenous knowledge creation for Asian economic development', *WZB Jahrbuch 1998*, 401–33.

Oakey, R.P. and T. White (1993), 'Business information and regional economic development: some conceptual observations', *Technovation*, **13**(3), 147–59.

OECD (1993), *Small and Medium-Sized Enterprises: Technology and Competitiveness*, Paris: Organisation for Economic Co-operation and Development.

OECD (1998), Issues paper for international conference on 'Building competitive regional economies: Upgrading knowledge and diffusing technology to small firms', Modena, 28–29 May.

OECD (1999), *Managing National Innovation Systems*, Paris: OECD.

OECD (2001), *Science, Technology and Industry Scoreboard 1999. Benchmarking Knowledge-based Economies*, Paris: OECD.

Patchell, J. (1993), 'From production systems to learning systems: lessons from Japan', *Environment and Planning A*, **25**, 797–815.

Pavitt, K., M. Robson and J. Townsend (1987), 'The size distribution of innovating firms in the UK: 1945–84', *Journal of Industrial Economics*, **45**, 297–306.

Pavitt, K., M. Robson and J. Townsend (1989) 'Technological accumulation, diversification and organisation in UK companies, 1945–83', *Management Science*, **35**, 81–99.

Pedler, M. et al. (1991), *The Learning Company*, London: McGraw-Hill.

Perroux, F. (1970), 'Note on the concept of growth poles', in D. McKee et al. (eds), *Regional Economics: Theory and Practice*, New York: Free Press, pp. 93–103.

Pike, F. and J. Tomeney (1999), 'The limits to localization in declining industrial regions? Trans-national corporations and economic development in Sedgefield Borough', *European Planning Studies*, **7** (4), 407–28.

Piore, M. and C. Sabel (1984), *The Second Industrial Divide: Possibilities for Prosperity*, New York: Basic Books.

Polanyi, M. (1996), *The Tacit Dimension*, London: Routledge.

Polt, W. (1997), 'Evaluation of technology and innovation policies – in search of best practices', *Platform Technology Evaluierung*, **5**, November.

Porter, M. (1990), *The Competitive Advantage of Nations*, London: Macmillan.

Porter, M. (1998), 'Clusters and the new economics of competition', *Harvard Business Review*, November–December, 77–90.

Pyke, F. (1994), *Small Firms, Technical Services and Inter-firm Cooperation*. Geneva: International Institute for Labour Studies.

Pyke F., W. Sengenberger (eds) (1992), *Industrial Districts and Local Economic Regeneration*, Geneva: International Institute for Labour Studies.

Reve, T. et al. (1992), *Et konkurransedyktig Norge*, Oslo: Tano.

Rizzoni, A. (1991), 'Technological innovation and small firms: a taxonomy', *International Small Business Journal*, 9 (3), 31–42.

Rosenfeld, S.A (1997), 'Bringing business clusters into the mainstream of economic development', *European Planning Studies*, **5** (1), 3–23.

Rothwell, R. (1983), 'Innovation and firm size: a case for dynamic complementarity; or, is small really so beautiful?', *Journal of General Management*, **8** (3).

Rothwell, R. (1991), 'External networking and innovation in small and medium-sized enterprises', *Technovation*, **11** (2), 93–112.

Rothwell, R. (1992), 'Successful industrial innovation: critical factors for the 1990s', *R&D Management*, **22**, 221–37.

RTP-Limburg (1996), *RTP-Limburg – Working Document*, Maastricht: Province of Limburg.

Sadler-Smith, E., I. Chaston and D.P. Spicer (1999), *Organisational Learning in Smaller Firms: An Empirical Perspective*, paper presented at 3rd International Conference on Organisational Learning, Lancaster University, UK 6–8 June.

Schmidt, E.M. (1995), *Betriebsgroesse, Beshäftigtenentwicklung und Entlohnung. Eine ekonometrische Analyse für die Bundesrepublik Deutschland*, Frankfurt, New York: Campus.

Scott, M. and P. Rosa (1996), 'Opinion: has firm level analysis reached its limits? Time for a rethink?', *International Small Business Journal*, **14**, 4, 81–9.

Seaton, R.A.F. and M. Cordey-Hayes (1993), 'The development and application of interactive models of industrial technology transfer', *Technovation*, **13** (1), 45–53.

Sengenberger, W., G. Loveman and M. Piore et al. (eds) (1990), *The Re-emergence of Small Enterprises. Industrial Restructuring in Industrialised Countries*, Geneva: International Institute for Labour Studies.

Smallbone, D. (1997) 'Selective targeting in SME policy: criteria and implementation issues' in D. Deakins, P. Jennings and C. Mason (eds), *Small Firms: Entrepreneurship in the 1990s*, London: Paul Chapman Publishing, pp. 127–40.

Smallbone, D., R. Leigh and D. North (1993), 'The use of external assistance by mature SMEs in the UK: some policy implications', *Entrepreneurship and Regional Development*, **5** (3), 279–95.

Smallbone D. and D. North (1998), 'Targeted support for new and young businesses: the case of North Yorkshire TEC's "Backing Winners" programme', *Journal of Small Business and Enterprise Development*, **5** (3), 199–207.

Smallbone, D., D. North, I. Vickers and I. McCarthy (1999), '*SME Policy and the Regional Dimension of Innovation: UK National Report. SMEPOL report no. 7*', CEEDR, Middlesex University.

Smith, K. (1994), 'New directions in research and technology policy: identifying the key issues', Oslo: *STEP Report*, 1.

Smith, K. (1997), 'Economic infrastructures and innovation systems', in C. Edquist (ed.), *Systems of Innovation. Technologies, Institutions and Organizations*, London: Pinter, pp. 86–106.

Smith, K. (1999), 'Industrial structure, technology intensity and growth: issues for policy', paper to DRUID conference on National Innovation Systems, Industrial Dynamics and Innovation Policy, Rebild, Denmark, June 9–12.

Soete, L. and A. Arundel (1993), *An Integrated Approach to European Innovation and Technology Diffusion Policy: a Maastricht Memorandum*, Luxembourg: Publication N° EUR 15090 of the European Commission.

Soskice, D. (1999), 'Divergent production regimes: coordinated and uncoordinated market economies in the 1980s and 1990s', in H. Kitschelt, P. Lange, G. Marks and J.D. Stephens (eds), *Continuity and Change in Compemporary Capitalism*, Cambridge: Cambridge University Press, pp. 101–34.

Stacey, R. D. (1996), *Strategic Management and Organisational Dynamics*, London: Pitman, pp. 21–49 and 62–4.

Stöhr, W. B. (1990) (ed.), *Global Challenge and Local Response. Initiatives for Economic Regeneration in Contemporary Europe*, London: Mansell.

Storey, D. (1988), 'The role of small and medium-sized enterprises in European job creation: key issues for policy and research', in M. Giaoutzi, P. Nijkamp and D. Storey (eds), *Small and Medium-Sized Enterprises and Regional Development*, London and New York: Routledge, pp. 140–62.

Storey, D. (1994), *Understanding the Small Business Sector*, London: Routledge.

Storey, D. and S. Johnson (1987), *Job Generation and Labour Market Change*, London: Macmillan.

Storey, D. and N. Sykes (1996), 'Uncertainty, Innovation and Management', in P. Burns and Dewhurst J. (eds), *Small Business and Entrepreneurship*, London: Macmillan, 2nd edn, pp. 115–37.

Storper, M. (1995), 'Regional technology coalitions: An essential dimension of national technology policy', *Research Policy*, 24, 895–911.

Storper, M. (1997), *The Regional World. Territorial Development in a Global Economy*, New York and London: Guilford Press.

Storper M. (1999), 'Regional economies as relational assets', paper presented to the International Conference 'Local Development in Europe: new paradigms and new schemes of economic policy', organised by Formas, Cattaneo University-Liuc, Insubria University, Varese, 1–2 October.

Storper, M. and B. Harrison (1991), 'Flexibility, hierarchy and regional development: the changing structure of industrial production systems and their forms of governance in the 1990s', *Research Policy*, **20**, 407–22.

Storper, M. and A. Scott (1995), 'The wealth of regions', *Futures*, **27** (5), 505–26.

Storper, M. and R. Walker (1989), *The Capitalist Imperative. Territory, Technology, and Industrial Growth*, New York: Basil Blackwell.

Teece, D.J., R. Rummelt, G. Dosi and S. Winter (1994), 'Understanding corporate coherence: theory and evidence', *Journal of Economic Behaviour and Organization*, 23, 1–30.

Tether, B. S., I.J. Smith, and A.T. Thwaites (1997), 'Smaller enterprises and innovation in the UK: the SPRU Innovations Database revisited', *Research Policy*, **2**, 19–32.

Tidd, J., J. Bessant, and K. Pavitt (1997), *Managing Innovation: Integrating Technological, Market and Organisational Change*, Chichester: John Wiley.

Tödtling, F. and A. Kaufmann (1999), 'Innovation systems in regions in Europe – a comparative perspective', *European Planning Studies*, **7/6**, 699–717.

Varaldo, R. and L. Ferrucci (1996), 'The evolutionary nature of the firm within industrial districts', *European Planning Studies*, **4** (1), 27–34.

Vázquez-Barquero, A., Javiez Alfonso Gil, Antonia Saez-Cala and Ana Isabel Viñas-Apaolaza (1999), *SME Policy and the Regional Dimension of Innovation: The Spanish Report. SMEPOL Report no. 6*, Madrid: GIDIT, Departamento de Estructura Económica y Economía del Desarrollo, Universidad Autónoma de Madrid.

Walker, W. (1993), 'National innovation systems: Britain', in R. Nelson (ed.), *National Innovation Systems. A Comparative Analysis*, New York, Oxford: Oxford University Press, pp. 158–91.

Weinstein, O. (1992), 'High technology and flexibility', in P. Cooke et al. (eds), *Towards global localisation*, London: UCL Press, pp. 19–38.

Wolters, A. and M. Hendriks (1997), *Monitoring Science and Technology Policy III*, Maastricht: MERIT.

Wyer, P. and G. Boocock (1996), 'The internationalisation of small and medium-sized enterprises: an organisational learning perspective', paper presented to the 7th ENDEC World Conference on Entrepreneurship, Singapore, 5–7th December.

Index

absorptive capacity 16–17, 20
additionality 140, 145, 151, 158, 162, 205, 206
Aproved Technical Service Institutes (GTS) 131, 149, 150, 152, 153, 154, 165, 182, 206
Apulia 65, 66, 71, 76, 83, 85, 90, 92–3, 101, 102, 103, 105–6, 121, 127, 142, 145, 149, 150, 207

BICs 204, 206, 208
broker 154, 156, 160, 208, 209
Business Link scheme (BL) 130, 155, 156, 157, 158, 159, 160, 165, 179, 180, 207, 208, 209

coherence
    aspects of 168
    bottom-up approach 168
    demand-side coherence 170
    external coherence, 170, 176, 178, 179, 189
    interactive perspective 175–6
    internal coherence 170, 174–5, 178, 179, 189
    linear perspective 175–6
    organizational coherence 171–3, 189–190
    regional aspect 180–81 see also spatial coherence
    spatial coherence 172
    supply-side coherence 170
    top-down approach 168, 173–4
coherent innovation policy system 171

Development Companies 141, 142, 145, 205 see also development corporations
development corporations 123, 206

endogenous (policy) strategy 64
ERP programme 124, 141, 142, 143, 144, 146, 147, 205
export specialization 59–62

FIRST enterprise 132, 161, 162, 163, 164, 207

governance
    direct 174
    indirect 174
Growth Fund 123, 141, 142, 145, 205

Hertfordshire 65, 73, 77

IMPIVA 126 see also technological institutes
Incubation and Technology Centre Wels (GTZ) 128, 149, 150
industrial district 27
Industrial Research Promotion Fund (FFF) 124, 141, 142, 143, 144, 205, 208
innovation
    definition of 10, 29
    barriers to 103, 218
Innovation and Technology Fund (ITF) 124, 141, 142, 143, 143, 145, 205
innovation behaviour 211
innovation centre 130
innovation costs 55–6
innovation policy 63–4, 88 see also innovation scheme
    central co-ordination 186–7
    definition 49–50, 78–9
    design of 209, 218
    for SMEs 113–4
    good practice 164–5, 204–9
    implication for 102
    main role for 198